Marginality, Migration and Education

Winniefridah Matsa

Marginality, Migration and Education

Educational Experiences of Migrants' Children in Zimbabwe

Springer

Winniefridah Matsa
Midlands State University
Gweru, Zimbabwe

ISBN 978-3-030-60872-9 ISBN 978-3-030-60873-6 (eBook)
https://doi.org/10.1007/978-3-030-60873-6

This Springer imprint is published by the registered company Springer Nature Switzerland AG
The registered company address is: Gewerbestrasse 11, 6330 Cham, Switzerland

Introduction

This book aims at exploring the nexus of marginality, migration and education in Zimbabwe. It provides a missing link between the topical issues of marginality, migration and education in the context of academic debates of Third World countries' various sources of marginalisation. The topical issues have a cause and effect relationship. Migration is not only recurring but is also increasing worldwide. People of South West Zimbabwe in particular, Bulilima and Mangwe, experience geo-physical, economic and socio-cultural marginality. This marginality results in the exodus of able-bodied parents to especially Botswana, South Africa, some abroad leaving children behind. Marginality can be a sign of discrimination, oppression or existence of a form of inequality. As such, it is a hindrance to the achievement of equity, equality and sustainable development goal number 4 which focuses on inclusive and quality education. Geo-physical and political marginalisation result in other various forms of exclusion, namely economic, social and cultural. The Bulilima and Mangwe people of Zimbabwe are experiencing political, educational, socio-economic and ethnic discrimination linked to geo-physical marginalisation.

Migration is one of the most crucial issues of our time and an important subject in communities worldwide. United Nations Department of Economic and Social Affairs notes that inequalities are reflected by figures of migration and that all types of migration are a result of force from civil, political, religious and economic reasons. International report by United Nations (2013) reveals that migration has continued to grow to unprecedented levels in recent years reaching 258 million in 2017 compared to 220 million in 2010. About 60% of migrants live in Asia and 80% in Europe. North America host 58 million followed by Africa at 25 million. Latin America and the Caribbean stand at 10 million. Since the international conference on population and development in 1994, the issue of international migration has risen steadily in the agenda of the international community. The 2030 Agenda for Sustainable Development includes several migration-related ratification instruments concerning international migration.

This book does not focus on all types of migration, but international labour migration. Nomadism and forced migration as a result of conflict across situations or internally displaced people as a result of natural disasters and wars are not part of the scope for this book. The book is about labour migration of people who make rational decisions and choices to migrate across borders. The focus is on migrants, emigrating, moving out as opposed to immigrants, which is entry into destination countries (immigration). It does not look at the life of migrants as immigrants in destination countries but their link with their left-behind families. It targets educational experiences of left-behind children due to labour migration across boarders (migrants' children). The book diverts attention from children migrating with parents to children left in sending countries. It delves into children's schooling experiences in the absence of parents, their relations with caregivers and teachers, as far as schooling is concerned.

The international community recognises that education is essential for the success of the 2030 Sustainable Development Goals. Sustainable Development (SD) Goal 4 aims at quality education and promotion of lifelong learning opportunities for all. In the World Forum in May 2015, UNESCO was tasked to lead and to coordinate the Education 2030 Agenda. The Education 2030 Framework for Action was also adopted in November 2015. Education is viewed as a means to transform lives of communities, reduce inequality and to wipe poverty. Education is also a basic human right which all humans have a right to particularly children (Article 28). Migration of parents leaving children in difficult circumstances for schooling hinders success of inclusiveness, equity and equality education. Lack of inclusiveness impacts negatively on the achievement of other sustainable development goals, namely eradication of poverty.

The book adds to existing scholarships on marginality, education and migration and expound on the nexus between the three. Migration is increasing and impacts negatively on the lives of children yet government and policy-makers seem to be blind to its effects, especially impacts on schooling and well-being of children left behind. It is also particularly important in Zimbabwe because the political and economic environment continues to push parents out of the country in search of greener pastures, yet migration policies and conditions do not allow migrants to take their families with them. As a result, the problems associated with migration of parents leaving families behind are likely to worsen. There is a large body of knowledge on education of children as immigrants in destination countries but very little on education of children left behind by migrating parents; hence, the book makes a strong case for this vulnerable group. There is also literature and policies on education and well-being of refugees and the displaced children in and outside conflict torn countries. The situation of children left behind by labour migrants is a forgotten one in migration and education policy issues.

The book has no interest in children as immigrants but exposes educational experiences of migrants' children in sending countries as given by children themselves (Child-centred approach). Children are not researchers themselves neither do they have voices in most researches about them. The book makes migrants' children visible and vocal in education and development as well as in research. It will help the governments, non-governmental organisations, teachers and communities to be aware of the situation of predicaments of migrants' children as a category of vulnerable children living and schooling in difficult circumstances. Who knows, the book might influence policy-makers to establish policies relating to migrant parents' leaving children of school-going age behind. The policies can assist children to cope with educational problems associated with the migration of parents. Above all, the book triggers further researches on the impact of migration on schooling of left-behind children in Zimbabwe and Southern Africa as a whole.

Gender has become part of development debates; hence, there is need to engender marginality, migration and education. While both genders experience marginalisation and discrimination in education, gender constructions make women and girls more vulnerable to the impacts of these. Experiences and severity of marginalisation and discrimination in education depends on one's gender (Gendered experiences). Impacts of migration also depend on gender. Implications to policy as a result need to take a gender eye into consideration. Although the book does not focus on gender per se, it reflects on gendered impacts of the intricate relationship of marginality, migration and education.

The structure of the book starts with exploration of the nature of international migration and the context of migration in Africa and Zimbabwe in particular as evidence of the fact that Africa and Zimbabwe have mobile populations. The chapter also gives a snapshot of patterns, reasons for and theoretical perspectives on labour migration. The book also determines the inter-connection between marginality, migration and education. Marginality, migration and education have cause–effect relationship. Marginality causes migration and migration impacts on education and further deepens marginality. Marginality also impacts on education as will be seen in Chap. 2. The gendered nature and impact of migration on care culture and family relationships as well as challenges on education are explored. Educational experiences of boys and girls relating to low-cost boarding, child renters, shifting care, strained and suspicious relations and neglect, salient wars between teachers and migrants' children are also discussed. The book concludes with discussion of western intervention strategies versus post-colonial funds of knowledge as well as synchronised coordinated model of policy and social strategies to help teachers and migrants' children cope with the impact of migration of parents in Zimbabwe. Implications of the book dovetail with strategies meant to assist teachers and migrants' children cope with schooling in the absence of parents due to migration.

In this book, literature review provides international and local contextual data on patterns of migration elsewhere, Zimbabwe, Bulilima and Mangwe laying the foundation for the magnitude, historical background, reasons for and theoretical perspectives on labour migration. It also provides information on geo-physical, socio-economic and political marginality and how these are intricate with migration to cause educational marginality in Zimbabwe. Literature review also exposes the gendered impact of migration and educational experiences not only in Zimbabwe but also elsewhere. It reveals effects of marginality and migration, educational challenges and highlights the risky physical and social environments resulting from marginality and migration. Literature expounds on children trapped in neglect, impact of migration on the family, care, relationships and children's rights. Finally, literature introduces debates on children's rights, equity intervention versus post-colonial strategies as context of existing and suggested strategies to help migrants' children cope with challenges of schooling in the context of absent parents. Information for the book is based on literature sources and primary data from a qualitative empirical study of Bulilima and Mangwe districts of Zimbabwe.

The empirical data was obtained from 250 participants who were teachers, learners who were migrants' children and caregivers. These were selected from six stratified schools, three from each district. The two districts had the highest number of migrants' children. Data collection was through focus group discussions with migrants' children in conventional and low-cost boarding schools as well as with child renting for schooling. In-depth interviews were held with class teachers and caregivers in the community of Bulilima and Mangwe. Questionnaires were administered to other teachers who were not class teachers and other migrants' children who were commuting from home. The study was child centred, aimed at unearthing educational experiences of migrants' children as given by children themselves, their teachers and their caregivers. Abramovich et al. (2011, p. 1) bemoaned the absence of child-centred methodology and child rights perspectives in researches and migration policies. The book then provides evidence-based experiences that can be used as basis for action in assisting migrants' children to cope with educational challenges emanating from absence of parents due to labour migration.

References

Abramovich, V Cernades P C, Morclachetti, A (2011) The rights of children, Youth and Women in the context of migration: Conceptual basis and principles for effective policies with human rights and gender based approach. UNICEF, New York.

UN (2013) International Migration Report. UN, New York.

Contents

Chapter 1
Background to the Nature of International Migration in the African Continent and Zimbabwe

International Migration

International migration flows have increased substantially over the decades. If migration increases at the same pace by 2050, the stock of international migrants worldwide could be as high as 405 million people. The number of international migrants is estimated to be 214 million people (International Labour Migration in Abramovich et al. 2011). International migration, according to United Nations in Raymer (2017), is movement to a country other than that of your own usual residence for a period of at least a year so that the country of destination effectively becomes your new country of residence. On the other hand, immigration is arriving in a new country which is not of your origin International Organisation for Migration (IOM) (2019). The book focuses on migrants' movement out of their country of origin (emigration) and not their entry or life experiences in their destination coutries. A migrant is a person who moves away from his or her place of usual residence, whether within a country or across an international boarder, temporarily. An emigrant moves out to a new country (IOM 2019). An emigrant and an immigrant are both migrants; however, the books look at migrants as emigrants, migrating parents leaving children behind in source countries.

International migration is now a global issue; the population migratory stock stood at 258 million in 2017. These are people residing in a country other than their country of birth making 3.4% of the world's population. About 5 million foreign-born persons entered the Organisation for Economic Cooperation and Development (OECD) countries in 2016 as migrant workers. Globally, $466 billion remittances were realised from low- and middle-income countries in 2016. About 68, 5 million individuals were forcibly displaced worldwide due to persecution, conflict, general violence, human rights violations or other reasons. There are 50 million irregular migrants estimated to be living around the world and 25, 4 million registered refugees were counted in 2017. About 102,800 refugees were admitted for resettlement worldwide (IOM 2019).

W. Matsa, *Marginality, Migration and Education*,
https://doi.org/10.1007/978-3-030-60873-6_1

About 6163 million migrants lost their lives or went missing during 2017 and 25 million of victims of forced labour were estimated in 2016. Out of those 5million may have crossed an international border. As a result of modern slavery, some migrants have been smuggled and trafficked for economic return of USD 5.5 to 7 billion in 2016. Children in 2017 represented 14% of the stork of international migrants and women made 48.8% of international migrants. About 18.8 million in 135 countries were newly displaced by sudden onset of disasters within their own countries in 2017, 13% of the adult population were potential migrants with plans to move permanently to another country in the next 12 months in 2015 and 75% applied for citizenship (IOM 2019).

The reasons for migration are classified within the theoretical perspectives of labour migration. The dual sector economy is one of the causes of outward migration in developing countries. The agricultural sector has excess labour and lower productivity to sustain families throughout the year. The industrial sector has high demand for labour; thus, migration is an economic decision. The dual sector economy results in the development of core and periphery centres where a periphery depends on the core (McCatty 2004). South African mines, for example, have more demand for labour than the traditional sector of peasant farming with no other economic alternatives to earn a living. This becomes an incentive for outward migration. South Africa is the core and other Southern African countries like Zimbabwe make up the periphery (Maphosa 2007). However, South African demand for labour has extended to other sectors of the economy where even educated professionals are absorbed in different jobs.

Neo-classical Marxists emphasise wage differentials between the area of origin and destination. Better wages in the destination become a pull factor. However, theories of new economics of labour reject structural forces in migration and argue that labour migration is a result of individual or family decision at household level who view migration as a risk management strategy (Konseiga 2005). The discussion of reasons for migration, in this chapter, borrows from all the theories and goes beyond to socio-cultural factors like the culture of migration.

Migration in the African continent

Migration patterns in Africa show that there is a huge wave of human movement. According to the bilateral migration matrix data of the World Bank, it was estimated that in 2010, the number of people living in countries other than their birthplace was around 215 million, of which 31 million were from Africa. The emigration stock for Africa was pegged at 30.6 million (Shimeles 2010). Africa is regarded as the continent with the most mobile populations. International migration in Africa accounts for at least 9% of the total global migrant stocks (Zanamwe and Devillard 2010).

In West Africa, both Ghana and Nigeria are sending and receiving countries. Ghana is a destination for migrants from Niger, Togo, Benin, Chad, Mali and Ivory Coast because of political instability and oil discoveries. Ghana also suffers from brain drain to the Far East, Dubai, China, Malaysia and other developed nations (Gunvor 2009). Migrants from war and conflict situations in North Africa migrate into Morocco from DRC, Liberia, Rwanda, Burundi and Sierra Leone (Adejopu 2004). The World Bank reported that 29 million emigrants from Africa in 2010 were from North Africa (Shimeles 2010).

In 2001, Democratic Republic of Congo (DRC) became a country of influx of immigrants into Central Africa due to a boom in the mining sector, and investors came from China, India, Korea and Pakistan. However, some Congolese people emigrated because of political instability. In East Africa, Sudan is the popular destination for Chad, Eritrea and Ethiopia while Kenya receives emigrants from Uganda and Tanzania (Adejopu 2004).

Migration in Southern Africa

Southern Africa has a long history of migration dating back to the mid-nineteenth century and South Africa is a popular destination with an economy which is a power-house attracting miners and other seekers of various jobs from far away places of central and East Africa. Emigrants into South Africa are from Botswana, Mozambique, Eswatini, Namibia, Lesotho, Zambia and Zimbabwe, among others. As a result of globalisation, distances by migrants have also increased. With the improved modes of transport, women and children have also become migrants. Migration patterns, trends, processes and complexities change as societies change (Shimeles 2010).

Since 1990, migration within Southern African Development Community (SADC) and from the rest of Africa to SADC has increased dramatically. The migrant stock of SADC was 3.4 million in 2000 and, of these, 2.4 million were from other SADC countries while 966 307 were from the rest of Africa (Crush and Williams 2010). The movement is characterised by labour migration that dates back to the colonial period. Labour migration from Southern Africa to South Africa was linked to the introduction of the capitalist mode of production which resulted in uneven development in the sub-continent. South African mining areas became pull centres. Beside economic reasons, the colonial period 1975–1978 experienced an exodus, resulting from activities of the liberation movements. Political refugees, detainees, rescrictees and freedom fighters migrated to South Africa, Mozambique, Zambia, Botswana, Tanzania and other Southern African countries that assisted their African brethren to attain independence (Maphosa 2004).

The employment of uneducated people and availability of unskilled labour as well as retrenchments in Lesotho increased the Basotho men's movement to South African gold mines, resulting in what Maphosa and Marojele (2013, p. 151) term the "culture of migration", where migration becomes an expectation and a normal part of life for a community. Zanamwe and Devillard (2010, p. 35) state that women have started migrating independently of men for wage labour, and they called this "feminisation of labour migration". Single men and women were found to migrate more than the married ones. However, married males compared to females migrated more because of societal expectations that the male should be the breadwinner. UNDP (2010) statistics reveal that out of all the men who migrated in 2007, 36% were single, 55% were married, 6% were divorced and 3% were widowed or separated.

In Botswana, engagement in underground South African mines proved to be a favourable option because the returns from mines were far much better than returns from agriculture. In Lesoto, in both source and destinations countries, there were negative consequences. There was distortion of the peasant economy, as well as a decline of agricultural economy making people to dependent on mining and mine wages.

The Problem and Context of Migration in Zimbabwe, Matabeleland South and Bulilima-Mangwe: History and Reasons for Migration 1890–2019

The characteristics of migration in Southern Africa and Zimbabwe mirror those of the region (Zanamwe and Devillard 2010). Zimbabwe has increasingly become a country of origin and, to a limited extent, a country of transition and destination. Emigration from Zimbabwe is not a new phenomenon, and labour migration to South African mining industries started as early as the beginning of the twentieth century. However, in the last decades migrants are employed in various sectors of the economy other than mining. Zimbabwe has historically been a country of origin and transit and continues to be a major source of migrant labour to Southern Africa and elsewhere (Zirima and Nyanga 2012; Zanamwe and Devillard 2010). This movement has largely been caused by the economic downturn which the country has been experiencing since the 1990s.

Maphosa and Marojele (2013, pp. 151–153) discovered that the culture of migration leads to low educational aspirations but “high affinity” for migration. Migration is a normal part of a life course representing transition from manhood and is widely accepted as a vehicle for economic mobility. Migration affects the value and increase the probability for future migration.

Crush et al. (2012) divide the phases into first, second and third wave of migration. The first wave started in the 1990s as a result of Economic Structural Adjustment Programmes (ESAPs) that saw many workers being retrenched from work to cut wage bills. The second wave was in 2002 to 2005, following the closure of industries and the seizure of commercial farms from white commercial farmers, resulting in more people losing formal jobs. The third wave started in 2005 to date being fuelled by both the economic and political crisis in Zimbabwe. Each wave had its own push-and-pull factors.

The colonial period was characterised by phase one with outward migration of political exiles, ex-detainees, refugees and freedom fighters to Mozambique,

Botswana, Tanzania and Zambia during the war of liberation in 1975 to 1978. In 1978, the United Nations (UN) High Commissioner for Refugees estimated that there were over 210,000 Zimbabwean refugees in these countries. Phase Two stretched from 1978 to 1989 and this was characterised by white Zimbabweans before and immediately after independence, fleeing military call-up and black independence that was imminent. Almost 50,000 to 60,000 whites left the country between 1980 and 1984 as a reaction of black independence (UNDP 2010).

Phase Three saw the post-independence conflict in Matabeleland and parts of the Midlands between the government and dissidents which led to a military operation which was commonly known as "Gukurawundi" from 1982 to 1987. This conflict fuelled more migrations. The estimated 4000 to 5000 refugees left for Botswana and South Africa (UNDP 2010). Phase Four (1990–1998) was a result of International Monetary Fund (IMF) and World Bank (WB) ESAPs in 1990, resulting in widespread hardships that influenced many professionals such as teachers, nurses, doctors and lecturers to leave the country in search of greener pastures abroad. This, Crush and Williams (2010, p. 5) and Tevera and Crush (2003), p. 3), call the "brain drain".

Phase Five (1999 to the present) saw the exodus of people as a result of Land Reform and redistribution programme, violence and the disputed 2000, 2008 and 2018 elections. Maphosa (2004, p. 6) noted that the 2000 Fast Track Land Reform Programme popularly known as "Jambanja", land seizure, was violent and was followed by a decline in production and an increase in unemployment. The largest number of emigrants was in the decade 1991 to 2000. This is echoed by Zanamwe and Devillard (2010) who add that Zimbabwe experienced a rise in emigration after 2000 most of which was illegal.

Most of illegal immigrants in South Africa and Botswana come from districts such as Bulilima, Mangwe, Gwanda, Beitbridge and Matobo in Matabeleland South, the hot spots of migration. These districts share borders with Botswana and South Africa. The local people in this area also share historical, kinship and linguistic ties with the Ndebele, Sotho, Venda, Tswana and the Kalanga which are spoken in the North, south of Limpopo and Ramokgwebana Rivers. The ties date back to the time of Mzilikazi's settlement in Matabeleland during the historic Mfecane movement of the Nguni generals who were fleeing from the wrath of Tshaka, the Zulu king. The geographical proximity and the availability of information about destination countries and accommodation, facilitate the movements (Maphosa 2004).

Beside historic reasons, there are also new economic, political, social and geo-climatic ones which push able-bodied people to seek employment elsewhere. Matabeleland South lies in a predominantly marginal and drought-prone region where agricultural production is mainly subsistence and rain-fed. Droughts and floods have increased the vulnerability of these people (Maphosa 2004). The region was declared a state of national disaster no less than four times between 1995 and 2001. The effects of ESAPs in the 1990s in Matabeleland South was exacerbated by recurrent droughts and lack of alternative livelihoods like gold panning which other people have resorted to in other parts of the country. The reasons for migration above can be summarised as geo-physical, economic, political, social and ethnic marginalisation.

With the land reform and redistribution during the (third Chimurenga), (Phase 6), coupled with its economic and political crisis from 2000 to date (2020), outward migration in Zimbabwe has increased in volume, spatial coverage and is no longer age, sex or race specific as before. It is non-discriminatory, the educated or uneducated, urban, rural, married or unmarried, formally or informally employed, legal or illegal, all find themselves crossing the borders to especially Botswana and South Africa. Political, geo-physical or social all converge to economic instability and unemployment which in turn force people to migrate elsewhere for employment outside Zimbabwe.

The other reason for migration noted by Maphosa and Marojele (2013) is the culture of migration, resulting from the community's experience of the benefits of migration. Some parts of Matabeleland region in Zimbabwe, migration to South Africa mark the coming of age of a young man. Those who have never been to South Africa are looked down upon. The culture of migration comes with peer pressure of "njiva" lifestyle (those who have been to South Africa). The Njivas become role models with all symbols of success which include cars, money, nice clothes, modern houses and technology, among others. This results in low priority given to education because the benefits of migration are perceived as better than those of education.

Some children drop or are forced to drop out of school by parents for non-payment of fees and wait for the time to migrate when they reach the majority status at 18 years. People of all educational levels migrate though those migrating for informal work tend to be more than the skilled ones. Zimbabweans in Johannesburg are relatively educated with 46% having completed secondary education, 22% had degrees and 38% had diplomas (UNDP 2010, p. 11). It is estimated that 0.6% of the population in Zimbabwe migrate abroad in search of employment (UNDP 2010, p. 90). Bulilima and Mangwe have been severely affected by migration. About 55% men and 45% women migrate to South Africa and Botswana in search of employment (Crush et al. 2012, p. 2). Zanamwe and Devillard (2010) highlight that migration results in movement of parents leaving children vulnerable. Madziva (2010, p. 125) echoes that migration of parents is "a disturbing phenomenon" hence the need to establish the impact of migration of parents on children left behind.

Trends and the Nature of Migration in Bulilima and Mangwe

The majority of teachers present in schools are diploma holders, most degreed and certificated teachers migrated to neighbouring countries during the 2008–2009 economic and political crises in Zimbabwe. Other teachers are either unqualified or temporary. Those with non-teaching degrees were laid off beginning of the year 2016 by the Ministry of Education in a bid to ensure quality education by qualified teachers with teaching degrees. The majority of current teachers have 3–10 years of experience. Teachers with above 10 years' experience migrated during the 2008–2009 economic crisis. The few remaining are administrators, heads of schools, deputies or senior teachers. Qualified and experienced teachers in Bulilima and Mangwe migrate more than the less-qualified and experienced. The culture of migration does not only affect the youths and adults of the community, but also the professionals. UNDP (2010) confirmed that Zimbabweans in Johannesburg were relatively well-educated, having completed secondary education and others are with degrees and diplomas.

The level of education for caregivers in the districts is low, a sizeable number have gone up to Ordinary Level while a few to tertiary institutions. Those with tertiary education are mostly teachers, civil servants working in townships or retired grandparents and relatives. A few have also reached standard 4 and 6. The rest are the elderly who have gone up to grade 7 and form 2; hence, children are left with the majority of caregivers who are less educated.

The rate of migration by youths to neighbouring countries is high. South Africa is the preferred destination followed by Botswana, the United Kingdom, Australia and the United States, respectively. Most of the learners (youths) migrate after completing Ordinary Levels or after failing to register for Ordinary Level Zimbabwe School Examination Council (ZIMSEC) examinations. Some of those who manage to write Ordinary Level examinations migrate immediately after writing examinations. Some do not bother to wait for the results while some never care to come back after release of the results. This high trend in migration is a miniature of what happens to adults of that community. Adult men and women also migrate in large numbers to South Africa, Botswana and overseas. In most cases, the ill and the elderly are found in the community villages. Children live alone, and most parents live outside.

Teenagers who drop out of school tend to migrate to Botswana first then later to South Africa. The advantage of Botswana is its proximity, socio-cultural and linguistic ties. Proximity makes transport costs cheaper and the youth can travel by foot. The distance from the last school in Mangwe district towards Mpoengs is 9 km. Madabe border post is also nearer. Some schools in Bulilima district are near Maitengwe border post and others, Ramogkwebane or Plumtree border post. They can also sneak across the border via paths undesignated as border posts.

Botswana is also a preferred destination because the Kalanga and the Tswana share historic, socio-cultural and linguistic networks as well as kinship ties. The language *isililima,* from which the name of the district Bulilima is derived, is a mixture of Kalanga and Tswana. They can also use English and Tswana because in some of these Zimbabwean schools, there are learners from Botswana whom they interact with and learn their language. The other reason why Botswana is the first option is because learners who drop out of school to migrate do it without the permission from migrant parents or caregivers hence in most cases they have no passports or pre-arranged accommodation in their destination. They cross borders on foot to nearby Tswana villages who take them for "piece" or temporary jobs without passports and offer them accommodation and food while they accumulate the pula for passports and transport to South Africa, the targeted destination.

Border jumping to Botswana is easier; youths can cross on foot, work illegally as cattle herders or domestic workers in Botswana villages. Most of those who migrate straight to South Africa are those who would have completed schooling or failed to write Ordinary Level Zimbabwe Schools Examination Council Examinations (ZIMSEC) as a result of failing to register by paying examination fees. These need permission from parents, migrant relatives or caregivers. They need transport and passports though some cross illegally through *omalayitsha*. *Omalayitsha* are cross-border transport operators who pirate the routes to and from Zimbabwe to South Africa through Beitbridge or Plumtree border posts via Botswana on a part-time or full-time basis. The term *malayitsha* is derived from the verb *ukulayitsha* meaning loading and transporting.

Omalayitsha transport passengers and goods on a weekly or monthly basis to and from South Africa to Zimbabwe legally and illegally. Illegal goods and passengers are expensive because the *malayitsha* drivers use clandestine means to cross borders with them. Sometimes an ox or cow maybe sold as transport cost for the illegal migrant, youth or adult. Most migrants from Bulilima and Mangwe prefer South Africa and Botswana as their destinations. Maphosa (2004) confirmed that most migrants found in Botswana and South Africa came from Bulilima, Mangwe and other parts of Matabeleland South in Zimbabwe because of proximity, economic, socio-cultural and linguistic ties.

Omalayitshas are seasonal or short-term migrants involved in transport cross-border trade. Their movement maybe monthly or fortnightly and have no intention of working in destination countries. Other cross-border traders also do not intend to settle permanently but may do piece jobs or concentrate on selling and buying of goods for trade. They use buses, private cars or omalayitshas as a mode of transport. Of late, some travel by planes. Cross-border trade is a common source of livelihood in Zimbabwe in the face of poverty and climate change. Like labour migration, it also cuts across all ages, ethnic group, religions and gender among others.

Bulilima and Mangwe migration has also altered the demographic population trends. More males than females migrate. Middle-aged and youths of 18 to 35 years dominate in migration, leaving a population of children and the elderly. The support of parents and teachers in education is lacking yet its crucial and practically indispensable. Children left with grandparents and other less-educated caregivers are disadvantaged when it comes to educational support (UNICEF 2008 in Delap 2013). Remittances from migration cannot teach children; hence, migration of the better-educated in Zimbabwe is "micro-level" brain drain affecting negatively left-behind children (Zirima and Nyanga 2012, p. 40). The movement of parents out of the country, leaving children behind, exposes them to new and hidden vulnerabilities.

The continued absence of close family members like parents disrupts child development and schooling. Children of migrants are perceived as "a disturbing phenomenon in current Zimbabwe" (Madziva 2010, p. 125). They are confronted with a "plethora of educational challenges" causing their educational performance to deteriorate (Zirima and Nyanga 2012, p. 40). This is worsened by the absence of child rights-based perspectives in migratory policies and legislation in countries of origin (Abramovich et al. 2011).

Conclusion

The African continent is characterised by mobile populations, as is the international community. Labour migration in Africa and Zimbabwe dates back to the period of colonialism and continues to increase. Migration in Matabeleland South, Bulilima and Mangwe mirror that of the regional and international phenomena. The culture of migration in Zimbabwe has spread its influence across youths, adults, uneducated, educated, highly qualified and experienced professionals, men and women, married, unmarried, legal and illegal migrants among others. Parents do not migrate with their children because of illegality of stay by some parents, types of informal jobs they do, accommodation expenses and migration policies in destination countries. The migration policies do not accommodate migrant families without workers' permits and legal documents for citizenship. Most children remain with extended family relatives, grandparents, caregivers or alone. Migration by parents for work purposes is an international phenomenon affecting children in sending and destination countries. However, the focus of the book is on left-behind children in Zimbabwe as a sending country.

References

Abramovich V, Cernades PC, Morclachetti A (2011) The rights of children, youth and women in the context of migration: conceptual basis and principles for effective policies with a human rights and gender based approach. UNICEF, New York

Adejopu A (2004) Recent trends in international migration in and from Africa. Human Resource Development Cent, Lagos

Crush J, Chikanda A, Tawodzera GC (2012) Third wave: Mixed migration from Zimbabwe to South Africa. SAMP and SARC, Cape Town

Crush J, Williams V (2010) Labour migration, trends and policies in Southern Africa: SAMP policy brief, Africa. SAMP, Queenstown

Delap E (2013) Protect my future: the links between child protection and population dynamics in the post 2015 development agenda. Available from save children 20130315 links. Accessed 11 Feb 2013

Gunvor J (2009) African perspectives on human mobility programme: comparative report on African migration trends. Oxford University: International Migration Institute, Oxford

IOM (2019) Glossary on migration IML series no. 34. Available from https://www.iom.int. Accessed 21 Oct 2019

Konseiga A (2005) Household decision as a survival strategy: the case of Burkina Faso. Discussion paper no.1819. Available from repec.iza.org. Accessed 9 Nov 2016

Madziva R (2010) "A living death": Zimbabwe migrants in the UK who are forced apart from their children. Nottingham: University of Nottingham. Available from http://ethses.nottingham.ac.uk. Accessed 2 Oct 2013

Maphosa F (2004) The impact of remittances from Zimbabweans working in South Africa and rural livelihoods in the Southern districts of Zimbabwe. Available from http://codesria.org. Accessed 29 Dec 2013

Maphosa F (2007) Remittances and development: the effects of migration to South Africa and rural livelihoods in Southern Zimbabwe. Routledge, London

Maphosa F, Marojele RN (2013) Changing the culture of migration: attitudes towards education among former Basotho labour migrants to South African mines. Afr Dev XXXVIII(1 and 2):151–151

McCatty D (2004) The process of rural-urban migration in developing countries. An honours essay submitted in fulfilment of the degree of Bachelor of Arts to Department of Economics. Carleton University, Ottawa. Available from www.dphu.org. Accessed 9 Nov 2016

Raymer J (2017) Measuring flows of international migration. https://wol.iza.org

Shimeles A (2010) Migration patterns, trends and policy issues in Africa. Tunis African Development Bank: Working Paper Series No. 119

Tevera DS, Crush J (2003) The new brain drain from Zimbabwe. SAMP Migration Policy Series No. 29. IDASA and Kingston/Queen's University, Cape Town/Canada.

UNDP (2010) Comprehensive economic recovery in Zimbabwe. Working Paper Series No 11 Harare: UNDP.

Zanamwe L, Devillard A (2010) Migration in Zimbabwe: a country profile 2009. ZIMSTAT and IOM, Harare

Zirima H, Nyanga T (2012) The cost of immobility: brain drain and educational outcomes of children in Zimbabwe. Wudpecker J Educ Res *1*(3):39–44. Available from http://www.wudpecker-researchjournals.org. Accessed 10 Oct 2013

Chapter 2
Geo-Physical and the Socio-Economic Learning Environments for Migrants' Children: Converging Experiences of Migration and Marginality on the Education of Boys and Girls in Bulilima and Mangwe

Geo-Physical and Socio-Economic Marginality of Zimbabwe, South West Zimbabwe and Bulilima and Mangwe (Zimbabwe)

Political, economic and social marginality of Zimbabwe can be traced back to its positionality of a third world country where the conditions of social and economic structures of non-western states are peripheral and dependent on global economy. Third world countries' socio-economic development and expansion is conditioned by another to which it is subjected to, resulting in the core and periphery relationship of exploitation (Smit 2013). This global dominance cause political powerlessness, poverty and social marginalisation. Although "third worldism" concept is dwindling in use, the former third world countries have failed to emancipate themselves from the core–periphery relationship. They have remained marginalised from active participation in the benefits of growth and social development as well as decision-making (Ayuba and Bakut 2014). Their agriculture, the backbone of their economies is firmly entrenched in the economies of the western world. The ecological marginality of Zimbabwe is superimposed on this global marginalisation.

Zimbabwe has agro-based economy and is a landlocked country depending on imports from manufacturing sectors outside the country. Agriculture is largely subsistence and rain-fed. Commercial agriculture employs a few and is heavily mechanised resulting in unemployment and economic migration to neighbouring countries. Agriculture is a key component for economic development accounting for 20% gross domestic product yet only 11% of the total national land area is arable. The rest of the land is marginal with vital land not suitable for cultivation. Drought or irregular rains occur impacting negatively on agricultural production forcing the country to rely on regional imports of crops like maize, cereals, tobacco and wheat. Maize, cereals and wheat deficit has increased over years. The drier regions of Southern and Western parts (South West Zimbabwe) have less arable land only appropriate for millet and sorghum yet maize is the stable crop in Zimbabwe. The risks of cyclical droughts are common in this rural part of the country.

W. Matsa, *Marginality, Migration and Education*,
https://doi.org/10.1007/978-3-030-60873-6_2

Agricultural production is also hindered by floods, pests, poor quality soils, lack of credit schemes and unresolved land tenure issues like the Fast track land reform since 2000 (USAID 2012).

About 75% of the population lives below the poverty line and rural poverty being highest in the provinces of Mashonaland central and Matabeleland. The population is concentrated in more arable and large urban areas of Harare and Bulawayo all depending on food from populated areas and less arable land. The land suitable for agriculture is in natural regions one, two and three. Natural region one has rainfall of above 1000 mm/year with specialised and diversified commercial agriculture but is the smallest with 2% land. Natural region two has 750–1000 mm/year with only 15% land appropriate for intensive farming. Natural region three has 650–800 mm/year of moderate and occasional dry spells translating into marginal growing areas. Natural region four has erratic and low rainfall of 450–650 mm/year suitable only for drought resistant crops but has 38% land. Region five where South West Zimbabwe, Bulilima and Mangwe are situated has less than 450 mm/year fit for cattle ranching and natural grazing but has 27% arable land. This is evidence that most of the land in Zimbabwe is not arable land and has food security risks (FEWS NET 2009). Currently, Zimbabwe is ranked as one of the poorest countries in the world, ranking at 172 out of 186. Weak economic performance relates to rain-fed-dependent sector, climate change-related factors, hyperinflation markets, governance, and land tenure system, lack of capital investment, trade deficits, national debts and corruption. Formal employment is estimated at 65% of national workforce (USAID 2012).

Evidence from Zimbabwe Poverty Atlas also concur that the two districts, Bulilima and Mangwe, are situated in the second poorest province of Matabeleland South. The poverty index for Bulilima, Mangwe in Matabeleland South stands at 73–84%. The area is worse off in terms of food security and nutrition, infrastructure utilities and social services despite sharing borders with Botswana and South Africa (ZIMSTAT 2015). The border posts Plumtree, Mphoengs, Maitengwe and Hingwe are entry areas for various goods but the poverty index still remains high. About 75% of the communal lands in the area are ecologically marginalised and uninhabitable (Hill 1999). The above becomes the source of marginalisation of the people of Bulilima and Mangwe. The able-bodied people then migrate to neighbouring countries in search of employment opportunities and alternative livelihoods. Circumstances force them to leave children or families behind. Policies in both sending and destination countries do not allow them to travel with children or their families. Most people migrate without work permits to search for any type of employment. Some take months before finding steady jobs.

Many Zimbabweans work informally within the country or in neighbouring countries. Formal sector unemployment puts pressure on household productive ages to migrate for better opportunities for employment in the diaspora and well–to-do cities of Johannesburg, Pretoria, Gaborone, Francistown, Lusaka, Livingstone, Maputo and abroad temporarily or permanently. It is estimated that 1–2 million Zimbabweans work in South Africa and other regional countries under varying degrees of permanence because of these geo-physical and socio-economic challenges (United Nations Development Programme (UNDP) (2013)). Most of these leave families and children behind. It is against this backdrop that this chapter explores how various forms of geo-physical and socio-economic marginality converge to impinge on the education of left-behind children. Their children remain experiencing different forms of marginalisation worsened by the absence of parents due to migration.

Marginality, poverty, inequality and exclusion are interlinked (Zhizhko 2016). Marginality is an underlying cause of poverty and inequalities. Inequalities are inherent in poverty. Poverty further marginalises the affected people. Inequalities also result in exclusion. Parents are then forced to migrate to neighbouring countries leaving children behind. The findings from Asia Research Institute confirm that Zimbabwe as a third world country experience economic marginalisation which gives rise to uneven development which encourages labour migration from rural to urban or developing to developed nations in search of employment (Yeah and Lam 2006).

Concepts and Types of Marginalisation

Marginalisation is a process of being pushed to the edges of a group, being given lower status or remotely considered (UNESCO 2010). As a result, the marginalised are pushed to harsh physical, economic and social environments, as are the people of Bulilima and Mangwe. Marginality, then, is being positioned on the fringes (Pelc 2006), the state of experience and being situated in the condition of exclusion and marginalisation. Marginality is a product of the process of marginalisation. It can also be a social phenomenon where a minority sub-group is excluded and have needs or desires being ignored. It is a process of putting individuals or groups outside mainstream society to form an acute and persistent disadvantage rooted in underlying social inequalities. Marginality is being discriminated against, excluded or having no right to access social services (Von Braun and Frank 2009). People experiencing marginality are prevented from enjoying equal opportunities and capacity development.

Marginalised people do not gain respect in their societies, are an "invisible minority" (Jenson 2000, p. 2). To be in marginality is not their choice but involuntary. The systems of marginalisation shape their destiny by preventing them from accessing resources and assets, services (Gatzweiller et al. 2011). This is the situation of migrants and their children. Systematic marginalisation is when political and economic systems generate inequalities through the distribution of social, political and economic benefits excluding others. Collateral marginalisation is a by-product of living among the marginalised where the state of being marginal reproduce itself, migrants' children live among the marginalised. There is also leverage where there is intentional pressure of demand of labour force in the developed nations influencing migration from countries of the global South to the North through unequal remuneration (Leimgruber 2004).

Marginality produces five forms of marginality, namely geometrics where groups or individuals are on the periphery of a larger area which can be state, continent or territory. There is also ecological one where the natural environment limits potential for survival excluding individuals or groups from benefits of the environment as we have seen above. Economic marginality is lack of access to production, access to resources and infrastructure to the extent of being on the periphery economically. Lastly, social marginalisation has to do with exclusion of minorities premised on language, religion, ethnicity and education among others (Leimgruber 2004). People in Bulilima and Mangwe experience all types of marginalisation. Marginality in Africa can be traced back to pre-colonial society and post-colonial interaction of leadership with its external actors. Colonialism was rooted in racial exclusion, exploitation and control of others. In post independent Africa, most states have been authoritarian and unable to redistribute economic benefits equally (Bond 2014).

The unequal development and marginalisation of Matabeleland South is colonial and political. Marginalisation and alienation of the peasantry in Zimbabwe dates back to 1890 British occupation when indigenous people were alienated from productive lands with the arrival of British colonial settlers. The indigenous people were pushed to unproductive marginal lands that later became the fringes of development (Hill 1999). The 1950s to the 1970s saw increased alienation of the blacks when the colonial settlers demarcated state farms and forest land through Land Apportionment Act 1931 and Land Husbandry Act 1951. What worsens the situation is that regions four and five of Zimbabwe have been historically disadvantaged and marginalised due to combined effects of ecological, colonial and political factors (CES IV 2003). Marginality, migration and education converge within this context to marginalise migrants' children in the community they live in. The relationship between marginality, migration and education is not a linear one (single arrow), one influences the other (Fig. 2.1). There is a cause and effect relationship, for example:

- **R1** Impact of marginality on migration—Geo-physical marginality causes inequality and poverty among people of South West Zimbabwe. Political

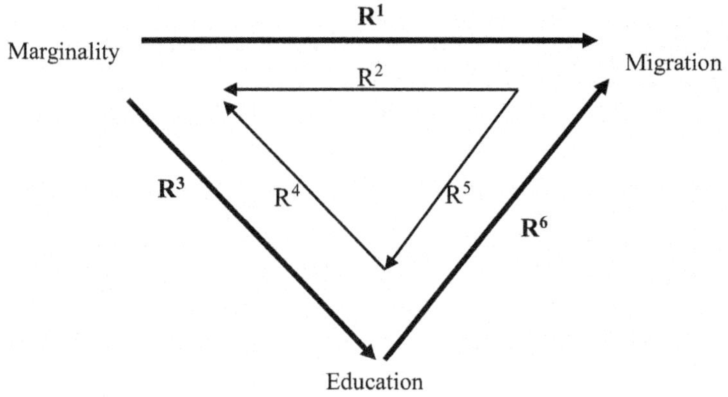

Fig 2.1 Cause and effect relationship between marginality, migration and education

marginality worsens inequalities and poverty at the same time causes exclusion from political voice, rights, privileges and access to resources. The distribution of resources is biased in favour of politically powerful regions and districts. Socio-cultural marginality also results in ethnic discrimination and marginalisation. Economic marginality converge with geo-physical, political and socio-cultural marginalities to become push factors for labour migration from South West Zimbabwe to countries in Southern Africa and abroad. South West Zimbabwe becomes the marginal periphery while South Africa and the countries abroad the core of development confirming the dual sector economy of core and periphery model by McCatty 2004). Maphosa (2005) highlighted that in Southern Africa, South Africa is the core and other Southern African countries including Zimbabwe make up the periphery. However, it must be noted that migrants from Zimbabwe do not migrate to South Africa alone but to other Southern African countries and overseas.

- **R2** Migration on marginality—The exodus of able-bodied people result in the marginalisation of the periphery of human labour causing low production in agro-based economy. As a result of poverty and food insecurity, the remittances are spent on livelihoods than development of the periphery. Migration causes drain on skills and knowledge (the brain drain) (Tevera and Crush 2003). South West Zimbabwe remains with a marginal population of the elderly, women and children who are not productive. This worsens the marginality of the region.
- **R3** Migration on education—Migration of the educated impacts negatively on education in the country of origin. The exodus of educated professionals drains the country or region of skills. Results of this study in Bulilima and Mangwe revealed that migration drained the two districts of educated and experienced degreed teachers leaving the inexperienced diploma and certificate holders, temporary teachers and those with non-teaching degrees. Migration is viewed as an alternative for livelihoods in a country with high rate of unemployment and low yields fuelling the culture of migration. The culture of migration causes low motivation for learning, low advancement in education and low retention in schools and increases dropping out of school to migrate.
- **R4** Education on migration—The unemployment of the educated in Zimbabwe fuels migration of the educated. High demand for the educated personnel elsewhere leads to the exodus of the highly educated and highly skilled. Education then becomes a pull factor for migration. Education also causes the educated able bodied to shun agro-economic production opting for white-collar jobs or other clandestine urban jobs.
- **R5** Marginality on education—Marginality has negative effects on education and lack of education further deepens marginality. Marginality results in poverty and unequal regional development. In South West Zimbabwe, marginality has affected educational infrastructure negatively. The area lack adequate schools and adequate teaching and learning resources. The schools are inaccessible because of distances of 16-20 km single journey. The roads are poor and rivers without bridges are a barrier. There are no learning centres like libraries, computer centres or museums as education support systems to ensure quality education.

Quality teachers shun rural schools because of poor roads, accommodation and general lack of educational resources. Low food production results in food insecurity, poor nutrition and poor health standards. Lack of quality teachers and resources lead to poor student performance, low priorities to learning, low retention and low advancement in education.

- **R6** Education on marginality—Education is a vehicle of development. It is supposed to eradicate poverty and inequalities. It provides communities with human capital skills for community development and knowledge for increased production. Lack of education or the educated cripples development and further marginalises the area educationally and economically.

Marginality, migration and education have this intricate and intertwined cause and effect relationship. Policy makers working on marginality, migration or education in this area should focus on this intricate relationship.

Ecological Marginality and Socio-Economic Learning Environments in Bulilima and Mangwe in Zimbabwe

The geography of the environment and its physical features has an influence on the learning environment of boys and girls. Learning environment refers to diverse physical locations, context and socio-economic and cultural context in which students learn (Thamarasseri 2017). To the teachers' perspective, the learning environment may be confined to the classroom content, learners' characteristics, skills, learners' support, resources and assessment. Learning environments, however, encompass the school, home, community, and the geo-physical, economic and socio-cultural settings. These influence learning directly or indirectly.

The educational infrastructure in Bulilima and Mangwe has traces of colonial and racial bottleneck system of education and neo-colonial inequalities of unequal distribution of resources by African governments. The educational infrastructure is lacking with a distance of at least 16–20 km with radius of 8–10 km between secondary schools and 8–10 km with radius of 4–5 km among primary schools. The distances are worse for newly resettled farmers following land reform and redistribution 1980–2000.

The roads are gravel with low and narrow bridges. In some villages, learners have to cross flooded rivers during the rainy season. This affects to a greater extent primary school learners and girls, their punctuality and attendance. These have to wait for the floods to subside or absent themselves because they cannot swim, climb trees or hold animal tails to cross rivers. The rivers in question are Tshangwa in the east of Bulilima, Nata (Manzamnyama) on the North and Thekwane on the South and West. Learners from Langabi (Kandana) area are forced to go to Gwambe than Ngwana and Matshinge because of Manheha River that has no bridge. Those who wish to go to Bezu secondary school have to cross Makobhodo stream and Tshangwa River. Learners from Masendu have to cross Thekwane River to Madlambudzi secondary school. In Mangwe, the rivers which are a barrier to schooling are Khahlu, Ngwizi, Sanzugwi and Simukwe. Learners from Linda to Mapulula primary school find it difficult to cross Khahlu River to school. Those from Mabulethi village and Kweneng primary school have to cross Ngwizi during the rainy season. Sanzugwi River affects those from Sangulube and Maninji primary schools and their communities. Simukwe River is also a problem for those from Kezi, Mambale area and Matokenye communities.

Learners board buses to Malalume and Mboma resettlement secondary school around 4 am to arrive at school before 5 am when schools start around 7.15 am. The distances are between 6 and 10 km a single journey. The delays in arrival have nothing to do with the distance but the buses are community ones ferrying commuters to town with their loads. They stop, pick and drop travellers along the way. Such environments are not gender and educational responsive. It makes learners vulnerable to sexual activities and love affairs at a tender age. In remote and newly resettled areas, learners also travel through thick forests or mountainous areas infested with wild pigs and leopards. Learners, who fail to be enrolled at Phumuza, travel through a thick and mountainous forest to Kandana (Langabi) school or to Bezu. This is not secure for boys and girls.

Some of the schools are located away from the big city of Bulawayo, about more than 180–200 km from Gampu, Madlambudzi in Bulilima or Mphoengs in Mangwe. The nearest town is Plumtree which is 203 km from Matshamhlophe, the furthest school in Mangwe and 120 km to Dupute the furthest in Bulilima. The district education offices are in Plumtree while the provincial offices are in Gwanda about 226 km from Plumtree. The national offices are in Harare the capital city which is about 540 km away from Plumtree. The distances from centralised and better educational facilities create a limiting environment in terms of schooling. Heads of schools do not easily access district and provincial education offices or communication documents on time.

There are no additional learning sites, for example, museums, libraries, learning centres, resort areas, community centres with Information or Computer Technology Centres yet education nowadays should be technologically driven. Kageler (2015) confirms that the infrastructure for ICT is limited, only 55% schools have electricity in the country, 1151 secondary schools have internet, 347 secondary schools have e-learning programmes with pupil-computer ratio of 404:1. The majority of these are private, mission, urban or town schools. Most rural day and satellite schools do not have such facilities. Heritage-based education encourages learning from historical sites, community libraries and museums for inter-generational knowledge and education for sustainable development. In Bulilima, only Luswingo cultural centre has since been established for heritage-based learning. There is none in Mangwe.

A few schools have computers donated, 18 in Bulilima and 16 in Mangwe out of a total of more than 80 primary and secondary schools. In addition the schools lack qualified teachers in computers and computer technicians to ensure consistent functioning of the donated computers. In each district, only 5 secondary schools out of 39 offer computer technology as a subject, and most of these are boarding schools. Distances to health centres or referral hospitals are long. The referral hospitals in the two districts are Plumtree and Brunarpeg with a distance of more than 170 km. Brunarpeg hospital serves all the schools and the communities in Mangwe. Bulilima has no referral hospital of its own but depends on Plumtree hospital.

Most of the schools have no buses or school trucks to ferry students in cases of emergencies. Local buses or commutor omnibuses travel very early in the morning once a day between 3 and 5 to Plumtree town or Bulawayo city. For the rest of the day, people depend on lifts which are not only unreliable but also not readily available. Most of the rural clinics do not have ambulances; cars are hired from teachers and neighbours even for ill learners and teachers.

There are three formal boarding schools in Mangwe and three low-cost boarding schools and three formal boarding. In Bulilima there are three low-cost boarding and three formal boarding schools. Distances between schools, few boarding and rural day secondary schools have given rise to sprouting of "bush boarding" accommodation and illegal rented boarding houses with poor social and sanitary facilities, risky and insecure for learners. Almost all day, secondary schools have learners living in rented houses in business centres, townships and illegal "bush boarding", some of which are squatter shelters with no institutional rules or supervision by parents or guardians. Learners experience marginalisation in terms of transport, health facilities, and accommodation as well as educational infrastructure. Some of the rural day schools do not have laboratories or agricultural workshops further excluding learners from science and practical subjects for projects, innovations and industrialisation being emphasised by the ministries of education in Zimbabwe.

The area is drought stricken with no piped water in most day secondary and primary schools. The community depends on river water, wells, dam water and a few boreholes. Tapped water is in towns, growth points, business centres and conventional boarding schools; otherwise, the majority of the community consume untreated water susceptible to diseases.

Socio-Cultural Marginalisation and Learning Environments

International migration increases in the context of the above marginalisation further deepening it. The two districts, Bulilima and Mangwe, lie in marginal boarder areas of South West Zimbabwe ravaged by poverty because there are no vibrant economic resources or activities as alternative for rain-fed agro-based agriculture which is doing badly owing to persistent drought. As a result, the two districts have high rates of migration to Botswana and South Africa. The place suffers from geo-physical and economic marginality which are major push factors for migration. Migration of youths and adults is high in the area owing to economic problems and the search for greener pastures.

In addition to the above, Bulilima and Mangwe are severely affected by what is termed "culture of migration" where migration becomes an expectation and a normal part of life (Maphosa and Marojele 2013). Although migration has benefits, it tends to marginalise education. School-going children drop out of school to migrate.

Educated parents and young men and women migrate leaving behind children with the less-educated caregivers who offer little or no support in schooling.

Socio-cultural marginalisation manifests in the education system through language of instruction and the dominant culture. Teachers and caregivers felt that although the schools are for the Kalanga people and some with Kalanga names, the dominant language and culture in schools is Ndebele. The language of instruction is Ndebele yet the local language is Kalanga. The Kalanga people are alienated from their own language and culture through education. Most of the teachers are trained in English and Ndebele as languages of instruction. However, since 2017, the government has since recognised the teaching of indigenous languages. Teachers for indigenous languages have been enrolled in teachers' colleges and universities for training. The languages are Kalanga, Tonga, Shangani, Nambiya and Sotho among others. Pilot schools have been set up in both Bulilima and Mangwe. This is a new development towards the right direction. Marginality resulting from migration is superimposed on these existing forms of marginalisation. Marginalisation then extends to all aspects of the life of the marginalised as noted by Zhizhko 2016. The marginal man internalise it and a cycle of marginalisation is reproduced.

Cultural marginalisation dates back to the arrival of Mzilikazi and the Ndebele who pushed the Karanga and the Kalanga to the marginal areas of Western Zimbabwe which is dry and unproductive. The Ndebele language and cultural hegemony were imposed on them. The system was firmly entrenched during the British colonial system of indirect rule where Ndebele-Nguni chiefs formally known as *Mzansi were* imposed as their rulers. Examples of Ndebele chiefs are Khandana (Langabi) Magutshwa, Gampo Maqhekeni Sithole and Mpukane Mpini Ndiweni in Bulilima. In Mangwe, Chief Wasi Ndiweni and the headmen Magcobafuthi Mdlongwa and Dimu Ngwenya among others. This is evidence of cultural hegemony, alienation and marginalisation.

Marginalisation of migrants and their children as seen is deepened by their historical background of marginalisation. Migrant parents, caregivers and the community marginalise education and view migration as a better option than education. Some learners have internalised that culture of migration pointed out by Maphosa and Marojele (2013) where migration is an expectation desired at the expense of education.

Conclusion

The chapter has explored the intricate relationship between marginality, migration and education. Geo-physical, socio-economic, historical background and socio-cultural reasons account for marginalisation of migrants' children's education and converge to push migrants' children to the periphery. These factors are a barrier to access to and quality of education. The infrastructure has been like this for the last two decades. No attempt is being made at policy level to look into the schooling environment of the learners in these rural areas.

References

Ayuba C, Bakut B (2014) Third world countries, marginal position in the contemporary global economy: an assessment. Available from SSRN https://ssrn.com. Accessed 16 Apr 2020
Bond P (2014) Marginalisation in Africa: is the end sight. Available from Ezine vol 15., http://www.africafiles.org/atissueezine. Accessed 10 Feb 2018
CES IV (2003) Sustainable development and natural resources management in Southern Zimbabwe. CES IV. Available from Italian cooperation. http://www.ester.it. Accessed 15 Feb 2018
FEWS NET (2009) Zimbabwe food security outlook, April to September 2009. FEWS NET, Washington, DC
Gatzweiller FW, Baumuller H, Landernburger C (2011) Marginality: addressing the root causes of poverty. ZEF working paper no 77. Centre for Development Research. Bonn.
Hill R (1999) Colonialism and inequality in Zimbabwe. IUCN-ROSA, Harare
Jenson J (2000) Thinking about marginalisation: what, who and why? Canadian Policy Research Networks, Ontario
Kageler S (2015) Education sector analysis Zimbabwe: final report. https://www.globalpartnership
Leimgruber W (2004) Between global, local marginality and marginal regions in the context of globalisation and deregulation. Ashgate, Aldershot
Maphosa, F. (2005).The impact of remittances from Zimbabweans working in South Africa on rural livelihoods in the Southern districts of Zimbabwe. Johannesburg: University of Witwatersrand
Maphosa, F. and Marojele , Marojele, R.N (2013) Changing the culture of migration: A attitudes towards education among the Basutho labour migrants to South African mines. African Development, XXXVIII (1 and 2),151-170
McCatty D (2004) The process of rural-urban migration in developing countries. An honours essay submitted in fulfilment of the degree of Bachelor of Arts to Department of Economics, Carleton University, Ottawa. Available from www.dphu.org. Accessed 9 Nov 2016
Pelc S (2006) Marginality and marginalisation.. Available at International pub. http://www.researchgate.net. Accessed 15 Feb 2018
Smit N (2013) The continued relevance of "Third world" concept.. Available from https://www.e-ir in for E-international relations. Accessed 16 Apr 2020
Thamarasseri, I. (2017) Gender, peace and education. Cambridge: A.P.H. Pub Corporation
Tevera DS, Crush J (2003) The new brain drain from Zimbabwe. SAMP Migration Policy Series No. 29. Cape Town/Canada, IDASA and Kingston/Queen's University.
UNESCO (2010) EFA Global monitoring report. Working group on call for EFA.
United Nations Human Development Report (2013)
USAID (2012) Best Zimbabwe report.. Available from https://www.usaidbest.org/docs/Zimbabwe Bellmon 2012.pdf. Accessed 15 Apr 2020
Von Braun J, Frank W (2009) Marginality: an overview and implications to policy. Centre for development, Bonn
Yeah BSA, Lam T (2006) The cost of immobility: Children left-behind and children who migrate with parents. Asia Research Institute: National University of Singapore, Singapore
Zhizhko EA, (2016) Socio-ecologically marginalised migrant farm workers in Mexico. Vol 5, issue 1(8) pp86-101
ZIMSTAT (2015) Small area poverty estimation: Statistics for poverty eradication. UNICEF and World Bank, Harare

Chapter 3
Gender and the Nature of Migration: Absence of One or Both Parents

Gender and International Cross-Border Migration in Zimbabwe: An Overview

International cross-border migration has a long and complex history and is unlikely to disappear (McDonald 2000). It remains an integral part of the social, political and economic fabric of countries in Southern Africa. International cross-border migration is directly linked to uneven development and socio-economic inequalities. It reflects global patterns of inequality and economic distribution of resources and benefits in original and destination countries (Giovanni et al. 2013). It is one of human kind s' oldest actions against poverty (Galbraith 1979). International cross-border migration is an attempt to overcome gaps between the rich and the poor member countries, regions or members of society. According to neo-classical economists, migration is a response to income differentials between regions or economic heterogeneity within developing countries, centre–periphery model suggested by dependency and under-development theorists (Faist 2016). Political power differentials between emigrating and immigrants regions also cause and manifest inequalities. Economic labour migration rises as a need to diversify income source for rural agricultural economies in the face of risks and crop failure as we have seen in South West Zimbabwe in chapter two. International border migration is one way to cope with risks by diversifying income through remittances.

International cross-border migration directly links migration, poverty, development as well as globalisation and transnationalisation. International cross-border migration by definition refers to mobility by international migrants from nation states crossing borders. It can also be conceptualised as flow of resources and persons in multiple directions across areas of emigration, immigration and onward migration to fight regional and social inequalities (Faist 2016). Social inequalities consist of uneven distribution of costs and benefits with respect to goods among individuals, groups, regions or states. Such goods maybe income, labour power, political, cultural and social capital. Inequalities are root causes of migration and

W. Matsa, *Marginality, Migration and Education*,
https://doi.org/10.1007/978-3-030-60873-6_3

both migration and inequalities result in differential life chances and opportunities although this may not be directly observable.

International cross-border migration by women is an attempt to bridge economic gender inequalities and an action against womens' poverty. However, it does not alter or transform power structures of inequality but help nations, regions and individuals to cope within these power structures. Migration of women is also a response to economic inequalities between men and women to alleviate feminised poverty (UNIFEM 2008). It is a response to gendered socially constructed economies which allow male access to formal education and employment. There is gender segregation of the labour force restricting women to the margins of the informal sector. As a result, migration of women has gradually shifted from family re-unification trend to a more economically motivated strategy (Faist 2016).

In economic labour migration, migration is an economic transaction between migrants and non-migrants (Taylor 1999). It is a risk diversification where income gains are maximised. Positive prospects are envisaged; as a result, both migrants and non-migrants may concur in decisions for migration. Decision-making for migration is not a domain of one person and is not always harmonious unless it is beneficial to both migrants and non-migrants family members (Nobles and McKelvey 2015). Bargaining theory by Agarwal (1997) and socialist–feminist arguments by Furguson in Bryson (1992) explain what happens between men and women in order for them to participate in the labour market in the public sphere. Agarwal (1997, p. 3) argue that decision-making is a form of bargaining especially between men and women. Intra-household interaction is characterised by both co-operation and conflict. Co-operation arrangement is reached when the outcome of co-operation is better than non-co-operation, when the outcomes are beneficial to both parties relative to non-co-operation.

Bargaining and negotiation of gender relations is very crucial in present-day societies where women are increasingly emerging in labour migration in a society which is both capitalistic and patriarchal. In a capitalist-patriarchal society, capitalist demand for labour contradicts patriarchal demands for labour and personal services in the home (Fergusen in Bryson 1992). Capitalism benefits from womens' labour in the public sphere and patriarchy benefits from control of womens' labour in the private sphere. Patriarchal men would be willing to allow women to join wage labour in the public sphere if it has more benefits than confining the woman to the private sphere. Joining or not joining the labour market for both men and women is an economic transaction bargained and negotiated. Winning or losing depends on the power of bargaining. Agarwal (1997) concur that gender division of labour is socially constructed and negotiated. For a woman to migrate, gender role activities like breadwinning, childcare, household chores as well as use of pooled resources are negotiated between men and women.

Nobles and McKelvey (2015) note that men and women can have differing views on the value of migration. Where the woman has control over household resources, the husband is likely to be allowed by the wife to migrate. Husband or wife can deny the spouse to migrate. Staying or migrating is a decision. In developing countries migration is a risk, so no member of the family act as a single entity when making migration decisions. However, decision to migrate maybe negotiated and agreed on or maybe imposed by the other spouse after failure of the bargaining process.

More and more women autonomously migrate to work abroad as independent household heads or family providers after the success or failure of negotiations. They may migrate alone leaving spouses and children behind. In some instances, both spouses migrate to the same or different countries as an economic strategy for survival leaving children alone or with caregivers. The nature of migration and its impacts depend on whether both parents have migrated or on the gender that has migrated and the gender of the child left behind. Sometimes the impacts have to do with conflicting views on migration, length of stay and extended family support.

Gender, Nature and the Impacts of Migration Elsewhere

In labour-sending countries, the growing numbers of children are left by one or two parents. Since 2000, 3000,000 children and adolescents have been left by one or both parents in Ecuador, Moldova, El Salvador, Jamaica and Albania. About 180,000 children aged between zero and 18 years in the Republic of Moldova have their parents working abroad. Of these children, 86,000 are left behind by one parent while 33,000 are left behind by both parents. About 31% of the 0–4 years have been left by one parent and 5.4% by both parents (UNICEF 2007).

In Bangladesh, between 14 % and 40% of all households, at least one member is working abroad. In Ecuador, about 200,000 boys and girls have at least one parent living abroad and the percentage of migrant women is higher than that of men (UNICEF 2008). In the Philippines, it is estimated that between 8.8 and 9 million children, around 27% of all minors in the land live separated from one or both parents (Parrenas 2010).

General statistics reveal that men migrate more than women do although this depends on the country's situation. In Sri Lanka, 1,000,000 children are left behind by mothers who migrate abroad in search of employment. About 13% of Mexican and almost 22% of Salvadoran immigrants, living in the United States have children left behind in the homes (UNICEF 2007). However, in Mexico, more men migrate to United States from the state of Zacatecas and many children grow up without the father. In Thailand and Vietnam, there is also more of male than female migration (Cortes 2007). In Jamaica, West Indies, more mothers migrate to the United States to work as nurses and nannies (Pottinger 2005). Also in the Philippines and Indonesia, women migrators exceed men (Graham and Jordan 2011). In many parts of Africa, including Zimbabwe, more males than females migrate to neighbouring countries (Zanamwe and Devillard 2010).

The absence of the father is understood differently because society tends to maintain traditional gender division of labour. The mother remains in the private sphere to do all the roles traditionally associated with women and the father can continue to display authority even from afar (Graham and Jordan 2011). Absence of the father results in increased household responsibility for women and children. In

Mexico, the challenges encountered as a result of the absentee fathers are those related to fatherly responsibilities, lack of protection and discipline. The absence of the father means more responsibilities for the mother and the children. The mother and the boy-child take over fatherly roles of provision and protection. In addition, the boy-child has to see to the issues of repairs of equipment and shelter, among others.

There is evidence of teen pregnancy, low school attendance and high dropout among girls because of lack of protection and discipline. With boys, there are challenges of alcohol and drug addiction, robbery cases, rudeness, drug-related deaths and imprisonment. Such kind of behaviour directly affects their school attendance, retention and achievement. Fathers have less contact with households than mothers; as a result, there is a communication gap with the children. This in particular affects the girls as left-behind children. As a result, in most cases fathers are not given the responsibility of caregiving if there is no elderly female to assist (Cortes 2007).

The gender of the migrant parent has different effects on the family and the child's security. The Health and Family Life Education Programme Evaluation in Sri Lanka reveals that 18% of girl-children, as young as 14 years old, who remained with fathers, have experienced forced sex as a result of lack of protection. However, male migration leave children better protected in terms of labour and income because the responsibility falls on the mother. Mothers may engage in income-earning projects to cover financial gaps of small remittances from the father (Baker et al. 2009).

In the Philippines, migration does not disrupt the development of children left behind, particularly if the mother remains behind (Cortes 2007). Children suffer the consequences of mothers' absence more than that of fathers. Migration of fathers causes boys to be burdened with work because they take charge of work outside the home. This is exacerbated by Islamic seclusion of women and girls.

The absence of mothers leads to violence or involvement in criminal activities in Trinidad and Tobago. About 80% of the children who had hearings in court had their mothers absent. These children also risk running away from home and act out behaviours so that they could be excused from lessons, tests, corrections and submission of homework or assignments. When the mother migrate, abuse, physical and emotional neglect are more likely to occur, especially to the girl-child. The boy-child is more vulnerable to physical abuse and neglect (Baker et al. 2009).

About 10.4% boys and 9.5% girls report physical abuse in the Philippines. Absence of mothers in the Philippines has a disruptive effect on schooling by lowering grades because of problems of social adjustment. The migration of mothers fuel worries and cause anxiety, leading to delinquency, profligacy, drug addiction and emotional frustration. Remaining fathers tend to rely more on the extended family support and children are often taken charge of by other family members rather than the father (Cortes 2007).

Most caregivers are grandmothers in Sri Lanka. Children with absentee mothers perform badly, especially girl-children who assume the role and household responsibilities of the migrant mother. About 20% of children of all ages show negative behaviour after their mothers' departure. Even if they have strong ties with caregivers, they still miss their mothers (Save The Children 2014).

In the districts of Colombo, Kurunugala and Gampala in Sri Lanka, in 2012, a total of almost 5000 females migrated abroad. Maternal migration has detrimental physical, psychological, educational and social impact on left-behind children. A sizeable number of fathers fail in sole parenting of children because of alcohol, drugs, extra-marital affairs and second marriages. The absence of mothers also leads to incest where fathers use young girls to satisfy their sexual needs. About 50% cases of incest are recorded in Sri Lanka where children between 15 and 18 years of age have experienced rape. Daughters become substitutes for migrating wives. The culprits mostly go unknown, unreported and unpunished.

In 2013, the Sri Lanka Bureau for Foreign Employment revealed frightening figures of child abuse in its report that covered 2009 to 2012. The highest was Kurunugala, Gampala and Colombo, respectively. The family dysfunction contributes to girls' poor social and psychological mental health, which in turn, affects performance and attitudes towards schooling (Perera and Rathnayaka 2013).

Men have poor adaptation to new roles in the absence of their wives. Society tends to blame migrating mothers for the dysfunction of the family and the suffering of children. Some children are exposed to sexual activities at an early age when fathers brought girlfriends into the small crowded homes of poor families. Those children born after the migration of the mother are sometimes discriminated against and regarded by fathers and relatives as illegitimate even if they are not. Fathers do very little to mitigate the effects of the mothers' absence. The breakdown of family functional roles in some instances, leads to early marriages, especially for girls. Both boys and girls are trapped in a cycle of hardships, abuse and uncertainty (Perera and Rathnayaka 2013).

The Gaza province of rural Southern Mozambique is characterised by massive male labour migration to South Africa and Namibia. Children whose fathers have been migrants for a long time have lower rates of discontinuation with schooling. There is unequal gender distribution of benefits from fathers' labour migration. Fathers' migration reduces boys' discontinuation with schooling rather than that of girls. Daughters are more likely to be pulled out of school to do household chores. However, children with highly educated mothers have lower rates of discontinuation with schooling compared to those with uneducated mothers. Lack of adult fathers did not only remove adult supervision over children but also increased stress for the remaining spouse. Financial benefits did not make up for the loss of fathers' physical presence (Yabiku and Agadjanian 2013).

Gender, Nature and Impact of Migration in Zimbabwe: Bulilima and Mangwe

Generally, in Zimbabwe there are fewer women compared to men who migrate for permanent employment outside the country. Patriarchal control does not allow automatically women to seek employment outside the home (Ranga 2015). Women also tend not to progress to degree levels and do not specialise yet outside there is demand for high educational levels and specialisation to meet critical skills deficits. Migration to the United Kingdom, for example, demands well-educated people with financial resources for travel and scholarships for furthering education. Most women in Zimbabwe do not possess the critical skills of mathematics, science and technical subjects and degrees which are on demand across borders. Educational levels of migrants are skewed towards males. This constrains lowly educated women from migrating.

Females migrating to South Africa also have challenges of accommodation and other vulnerabilities, like sexual harassment in their childcare roles in the private sphere employment. The total number of women migrants into South Africa from Zimbabwe in 2005 was 44%. The range of immigrants in terms of age, occupation and educational levels has become diverse. Womens' gender roles have shifted from child nurturing to provision of financial support. In the 1990s, of the Zimbabweans who migrated to South Africa, 39% were women mostly household heads and primary breadwinners (Crush et al. 2012).

Female migration increased following economic structural adjustment programmes in the 1990s and 1999 land reform, but female migration by the educated females still remains low compared to men. There are more females than men in cross-border trade than in labour migration for permanent employment. However, migration to the United Kingdom saw more females of the moderately educated. About 52% Zimbabwean females have the advantage of their caring roles of nursing in the health sector and social work in care institutions of children, the elderly and the disabled (Moyo and Peruma 2018). The above shows that, generally in Zimbabwe, females are not as educated as males. The educated males migrate more than the few educated females. Sometimes both parents migrate leaving children with illiterate; semi-educated or moderately educated female and male caregivers. Migration takes away one or both parents and the better educated in general. The impact of migration on children depends on whether one or both parents migrate, the levels of education of left-behind caregivers and the gender of the parent who migrates.

In Masvingo town of Zimbabwe, the absence of both parents between February 2010 and October 2011 had adverse effects than absence of one parent. Thirteen months after the departure of both parents, performance of children dropped because of changes in family composition and roles that affected homework and schoolwork. Children with absentee fathers were seriously dropping out of school compared to those with absentee mothers only. Children with absent mothers, especially girls, drop out of school to help with household chores; as a result, they have poor attendance. Absence of the mother changes the living conditions and jeopardises educational achievement. Children raised by one parent often face educational problems, especially if the male parent is rearing a female child and a female parent rearing a male child (Zirima and Nyanga 2012).

Children are affected more by the absence of the parent of their own gender. Generally, single parents have less time to help children with homework, are not consistent with discipline and have less parental control over the child of an opposite sex. The parent of the opposite sex is less likely to fully assist the child on emotional issues that affect educational performance. However, the effect depends on the individual child characteristics, household characteristics and the caregivers (Zirima and Nyanga 2012).

Migration alters the demographic population trends in Zimbabwe. More males than females of middle-aged and youths of 18 to 42 years dominate migration, leaving a population of children and the elderly. More males than females migrate because society still believes in male breadwinning. Most children remain with their mothers or grandparents, especially grandmothers. Fathers are rarely given responsibilities to look after children if there is no female to assist. Fathers are not the best option for caregiving to migrants' children. Female migration and caregiving by fathers results in domestic upheavals. Children left with fathers cannot cope with absence of mothers. They are deprived of attachment and they experience loneliness and emotional stress.

A sizeable number in Bulilima and Mangwe remain with employed caregivers while some stay in child-headed families with other siblings. Sisters assume motherly roles while brothers take up fatherly roles. A few live with foster guardians who are mostly females. Foster guardians include family friends, teachers living in cottages, property owners who own rented houses and other known community members living near the schools. In all these residences, there are more females to care for the children than males. The people in this community believe in "feminisation of care". Men are not trusted when it comes to care of children.

Migrant parents and their children prefer mothers and grandmothers as the best options for caregiving. However, there are some who are left with aunts, uncles (*malume*) and uncles' wives (*malumekazi*), employees and foster parents. Of interest is that the definition of uncles (*basekulu/malume*) in a Kalanga and Ndebele culture includes relatives from the mothers' side, and aunts (*bamadzi/babakazi*) are those from the fathers' side. Generally, relatives from the mothers' side are preferred. These are grandparents from the mother's side, grandmother, mother's sisters, mother's brothers. An uncle in a Ndebele or Kalanga culture gets into the shoes of the mother and with assistance of his wife is the rightful person to look after the sisters' children. Where bride price has not been paid by the father of the child, an uncle can use his powers to claim care for the child or even refuse to give the child to his brother-in-law. Children in child-headed households stay with their brothers and sisters of almost the same age or can have hired workers. In most cases, boys are the ones who stay with headboys and girls stay mostly with hired elderly female workers.

In Bulilima and Mangwe, there are more females than male teachers. Male teachers migrate more than females. Other contributing factors are that generally, the national statistics has more females than males and the teaching profession is also feminised. More males than females with better qualifications and experience have migrated. The majority of male and female teachers had 3–10 years of experience. Male teachers with above 10 years migrated to South Africa and Botswana during the 2008–2009 economic crisis. Unfortunately, because of the patriarchal nature of society, better qualifications for males and the experience as a result of historical advantages enjoyed by males, the two districts have a higher number of males in administrative posts. Five out of thirty nine heads of schools are male heads. Males get frustrated by the political and the economic crisis and take the risks of migration.

The level of education for caregivers in both districts is low, a sizeable number have gone up to Ordinary Level while a few have been to tertiary institutions. Males are better educated than female caregivers considering the fact that in this area there are more female caregivers and that generally the level of education for females is lower than that of males. Caregivers with tertiary education are mostly teachers staying with migrants' children in cottages or in their neighbouring homes, civil servants working in townships or retired grandparents and other relatives. A few have also reached standard 4 and 6 and the rest are the elderly who have gone up to grade 7 and form 2. The majority of children are left with caregivers who are not or are less educated.

The illiteracy of female was also noted by Ganju (2012) that two thirds of illiterate people around the world are women. ZIM Statistics report that in Matabeleland South where Bulilima and Mangwe are situated, literacy rates for both males and females are low (Zanamwe and Devillard 2010). Maphosa (2004) also highlighted that this part of the country has very few people who proceed beyond secondary level and the completion rates for secondary education are low because of the culture of migration. The majority of caregivers who are grandparents, wives of migrants, house cleaners, herders and other relatives have low literacy levels. As a result, they are unlikely to assist migrants' children in educational issues.

Migrants' children whose fathers are migrants exceed those whose mothers are migrants. Male youths migrate more and have aspirations to migrate than females. Learners and youths migrate after completing Ordinary Levels or after failing to register for Ordinary Level Zimbabwe School Examination Council (ZIMSEC) examinations, to greater extent males. The reason why more male youth migrate immediately after writing examinations is because, in most cases, males do not bother to get permission to travel. They are not worried about security issues or places of residence at their destination countries.

Boys who would not have completed schooling tend to migrate to Botswana first to work as cattle headers, garden boys, cooks in canteens or as a butcher, then later to South Africa. Girls do domestic work or work in flea markets. Some girls fail to proceed because of marriage and lack of passports. Boys cross the borders without passports or without transport. More males than females are border jumpers, crossing the boundaries illegally sometimes by foot. This is popularly known as "*dabulaphu*" or border jumping using foot as a mode of transport. Botswana is nearer and both males and females can travel by foot though males do it more often than females.

The other reason why Botswana is the first option is because both males and females can fit well in Botswana even through marriage owing to socio-cultural and linguistic networks as well as kinship ties. Learners who drop out of school to migrate do it without the permission from migrant parents or caregivers; hence, in most cases they have no passports or pre-arranged accommodation in their destination countries. Some cross illegally through *omalayitsha*. Kinship and linguistics ties gives them an advantage.

Omalayitsha are very expensive and the trade is dominated by male transporters. Parents fear for the security of girls since the *omalayitshas* in most cases use clandestine means of crossing the borders. Parents sometimes need to sell an ox or cow as transport cost for the illegal migrant, youth or adult. This can only happen if permission has been granted by parents, caregivers and migrant parents.

More female learners than male ones are left in mission and low-cost boarding. In two of the schools in Bulilima and Mangwe, there were 90 boys and 136 girls, 67 boys and 108 girls in low-cost boardings in 2016, respectively. Most of the caregivers are middle-aged between the ages 41–59 and 60 plus years. For teenage boys, they range between 22 and 50 years. Female migrants' children are left with mostly elderly caregivers who offer little assistance as far as education is concerned. The ages of children cared for by the female caregivers range from 0 to 19 years at most. However, for those left at primary and secondary schools, the dominant years are 10–18. Those above 18 years once migrated with parents and returned to repeat or have been in Zimbabwe but repeated a level of education. Most of these children, especially boys, live in child-headed household or live with employees.

The decision to migrate by parents in most cases is not discussed with all members of the family. Only the two parents (mother and father) discussed this. Sometimes, they discuss with other members of the extended family excluding children, especially if it is the father who is migrating. On average, most of the children are too young to participate in the discussions but even the older ones are informed of the decisions after the father has migrated. Male migrants migrate even if the idea of migration is opposed by their wives and other family members, sometimes they can migrate without informing anyone only to inform family members, including the wife, when they arrive at their destinations. The intentions to migrate by female members, their destinations and place of residence on arrival are made known well in advance. This is because in patriarchal families, males' decisions are more independent than that of female.

The length of stay by parents depends on gender and availability of travel documents. Male length of stay range from one year to twice a year or can return at retirement and the intervals for visits are spaced. Female stay is shorter and the visits are more often. Of interest to note, is the fact that migration by parents takes place at any time during the child's lifetime. Some children have always known their fathers as migrants if the father migrated before they were born or when they were very young. They have never stayed with them. A child can complete primary schooling living with the mother or caregivers. Such children have always known their fathers as visitors and some who are left when they are toddlers or before birth are introduced to their fathers.

Some children born in the diaspora are brought to Zimbabwe to begin pre-school level. These remain under the custody of the caregivers, especially grandmothers throughout their years of schooling. However, foster parents, friends and employees tend to be given children for care towards and during secondary level. The frequency of visits varies from once a year, twice a year to once in two years depending on gender. The majority of migrant fathers visit their left-behind families once a year during the Christmas holidays in December. Some mothers come twice a year, during Easter and December holidays. Some fathers visit once in two years while others do not visit at all. The length of migration and frequency of visits show that there are no close ties between migrants and their children. Children can spend their childhood or teenage years being with caregivers until they migrate as young adults.

On average, there is frequent communication with mothers than fathers. It is rare for children not to communicate with their mothers. Mothers also communicate more with teachers and caregivers. Fathers communicate with teachers once per term in relation to performance or fees. The few fathers who communicate weekly, fortnightly or monthly rarely communicate with their children directly but with the caregivers if there is an issue to be settled. It is up to the caregiver to suggest to the parent to talk to the child. Children prefer communicating with mothers than fathers. They also prefer that teachers talk to mothers than fathers. This shows distance between male migrants and their children.

Crush et al. (2012) discovered that generally in Zimbabwe women maintained strong ties with children and kins. They provided better long distance guidance and financial support. They send money and interact regularly with kins. They are selfless and not selfish. They are respected for playing proper supportive female roles. Female migration is viewed, as benefiting kins more compared to males. They are better in extended family kin keeping, fulfilling their roles of caregiving. Females respond quickly when children and kin needs arise. Their roles have shifted from resident to economic caregiving and kin keeping. They tend to care for the extended family as a whole (extended and generational ties) while men concentrate on their own nuclear families.

Tevera and Chikanda (2009) also discovered that communication by women was more frequent in spite of tremendous distance thereby providing emotional care and guidance. Women have a sense of responsibility than men hence more attractive as migrants. Women talk on phones are lengthy, more frequent, often on a daily basis. They text or visit regularly and promise to return at retirement. Women have managed to shift from direct to indirect care. They no longer nurture children at close proximity but practice distance nurturing referred to as "transnational motherhood" in the Philippines where the concept of mothering is expanded to include "breadwinning" (Parrenas 2010, p. 1828). However, financial mothering is not sufficient to groom and support children in education.

In Bulilima and Mangwe, more boys than girls have a desire to migrate. The attraction for remittances like cars, expensive cell phones, iPads, bicycles, fancy clothing and jewellery influenced their desire to migrate. Girls are attracted to clothing, jewellery, toiletry and they do not mind whether these are brought by parents, relatives or boyfriends. Fewer females drop school for migration, sometimes they drop to get married to *njivas*.

Both show symptoms of lack of father figures to a greater extent though grandmothers and mothers' sisters seem to cover the gap for absent mothers. Both are not comfortable with male relatives. Male relatives make it clear that children are foster ones and not their offsprings. Absence of fathers has negative effects on boys than girls. Boys lack guidance and role models. The boys downplay advices given by relatives. They absent themselves from lessons. They do not write homework or work given in class. They are reckless and always start work on new books every time. They spend large sums of money at shopping centres during break and lunch times. This is evidence of lack of budgeting skills supposed to be learnt from parents. Some have pocket money of up to R 100 to spend a day.

The above is consistent with findings by Kufakurinani (2014) who discovered that in Masvingo, Zimbabwe, mothers left behind by migrant spouses do not fill the void for male authority. Boy children lacked male encouragement, control and role modelling. Boys, because of lack of male authority, squandered fees with friends and girlfriends. Others developed truant behaviours.

During the holidays, boys assist their parents as *malayitsha* conductors and some eventually drop to be *omalayitshas* themselves. They are difficult to control by both male and female relatives at home and by teachers at school. Some are aggressive, rude, bully, extravagant, unloving and emotionally unstable. As a result, some boys take to drinking, truancy, outgoing with gangsters, in serious cases, theft, robbery, drug abuse and fist fighting. Absence of parents especially the father has detrimental effects on social and psychological development of boy children. Female learners affected by the absence of parents manifest loneliness, confusion, isolation and emotional instability. Absence of the mother disturbs girl development and management of womanhood. They have no one to talk to about issues relating to sexual maturation and reproductive health save for those staying with grandmothers, aunts and mothers' sisters.

Conclusion

The nature of migration in Bulilima and Mangwe, Zimbabwe, is gendered, so is care, decision to migrate and the impacts of migration. Reasons for migration, mode of travel, length of stay, frequency of visits, communication and decisions to migrate all depend on gender. Caregiving is feminised thereby disadvantaging boy children. The gap for a father's absence is difficult to fill than that of a mother. As a result, there is need for migratory policies in both sending and receiving countries to have a gender eye in order to curb gendered impacts. Existing migration policies are gender neutral assuming that migration affects men and women the same way.

References

Agarwal B (1997) "Bargaining" and gender relations: within and beyond the household. Femnist Econ 3(1):1–51

Baker C et al (2009) The impact of migration on children in the Caribbean. Available from http://www.unicef.org. Accessed 23 Jan 2014

Bryson V (1992) Feminist political theory. Basington: The MacMillan press
Cortes R (2007) Children and women left behind in labour-sending countries: an appraisal of social risks. UNICEF, New York
Crush J et al (2012) Third wave: mixed migration from Zimbabwe to South Africa. SAMP. Migration policy series no 59.
Faist T (2016) Cross-border migration and social inequalities. Annu Rev Sociol 42:323–346
Galbraith JK (1979). The nature of mass poverty, Mass. Cambridge: Harvard University Press
Ganju E (2012). A mothers' day tribute: the role of women and parenting in girls' education around the world. Available from http://www.huffinton post.com/ganju. Accessed 21 June 2013.
Giovanni J et al (2013) A global view of cross-border migration. NBER Working paper no 16646. Available from https://joxeu.org
Graham ES, Jordan LP (2011) Migrant parents and psychological well-being of left-behind children in SE Asia. J Marriage Fam 73(4):763–787
Kufakurinani U (2014) Transnational parenting and emergence of diaspora orphans in Zimbabwe. Brill: Africa Diaspora, Leiden
Maphosa F (2004) The impact of remittances from Zimbabweans working in South Africa and rural livelihoods in the Southern districts of Zimbabwe.
McDonald DA (2000) Towards a better understanding of cross-border migration in Southern Africa. SAMP. St Martin s' Press, New York
Moyo Z, Peruma J (2018) Globalisation and experiences of Zimbabwe female migrant teachers. J Educ Stud 17:1. Available from https://www.researcgate.net. Accessed 19 Apr 2020
Nobles J, McKelvey C (2015) Gender, power and emigration from Mexico. Demography 52(5):1573–1600
Parrenas RS (2010) Transnational mothering: a source of gender conflicts in the family. North Carolina Law Rev 2010:1825–1856
Perera V, Rathnayaka MR (2013) Sri Lankas' missing mothers. A working paper on the effects of mother migration on children. Save the Children, Baticoloa
Ranga D (2015) Gender differences in the migration of teachers to South Africa. East Afr Soc Sci Rev 31(1):43–62
Save the children (2014) Left behind, left out: The impact pdf children and families of mothers migrating for work abroad. Summary Report. https://resourcecentre.savethechildren.net./Library/left-behind-left-out-impact-children-and-families-mothers
Taylor EJ (1999) The new economics of labour migration and the role of remittances in the migration process. Int Migr 37(1):63–88
Tevera D, Chikanda A (2009) Migration remittances and household survival in Zimbabwe. SAMP. Migration policy series no 51
UNICEF (2007) The impact of international migration: Children left-behind in selected countries of L. America and the Caribbean. UNICEF. Available from www.unifem.org. Accessed 17 Jan 2014.
UNICEF (2008) The effects of parental deprivation on the development of the child left behind by Moldovan migrants. UNDP/CRIC, New York
UNIFEM (2008) Unleashing the potential of women in informal cross-border traders to transform intra-African trade. Available from www.unifem.org. Accessed 14 Apr 2020.
Yabiku ST, Agadjanian V (2013) Men's labour migration and schooling of children left-behind in rural Mozambique. Arizona State University, Centre for Population Dynamics, Tempe
Zanamwe L, Devillard A (2010) Migration in Zimbabwe: a country profile 2009. ZIMSTAT and IOM, Harare
Zirima H, Nyanga T (2012) The cost of immobility: brain drain and educational outcomes of children in Zimbabwe. Wudpecker J Educ Res 1(3):39–44

Chapter 4
Effects of Migration Elsewhere and Zimbabwe

Effects of Migration

Although migration of parents has negative impacts related to the care and well-being of children left behind, it has positive effects that result from remittances for schooling. In El Salvador, Edward and Ureta (2003) discovered that enrolments among 10–17 years correlated with international migrant monthly per capita remittances. There was high school retention rate in a country with 30% adult illiteracy among 40–50 years. Outward migration enabled parents to finance both private and public education. About 14% rural and 15% urban households received remittances from abroad. The years 1980–1983 enjoyed the largest improvement in school enrolments and retention. About 40% of school-going children with migrant parents completed primary education. Enrolments reached the equivalent of universal primary average of 80% rural and 90% of urban schools.

Migrant remittances can alleviate household budget constraints, allowing parents to afford school fees and other educational expenses. Remittances can also reduce economic shocks and are a strategy for poverty reduction in the source country. It is also a mechanism to earn attractive wages. In the destination country, migration alleviates labour shortages and boost skills. This was observed by Federici and Gannett (2006), Oropesa and Robles (2011) in Peru and Lima.

However, in the source country, migration may create labour shortages, drain skills from the economy and cause disintegration of families, while also flooding the labour market in the destination country, leading to unemployment and illegal activities (Maphosa 2004). In Peru and Lima, migration is beneficial to schooling. There is a strong association between schooling and remittances. The increase in remittances increases the likelihood of enrolment in both urban and rural areas. Those households receiving remittances also have high retention rate in schooling.

In the Philippines, remittances increased access to education and retention in schooling. They afforded parents chance to send their children to private schools which had better resources than public ones. The level of education increased

W. Matsa, *Marginality, Migration and Education*,
https://doi.org/10.1007/978-3-030-60873-6_4

thereby delaying entry into labour force (Rossi 2008). In Romania, migrants' children achieved higher grades and were motivated to learn knowing that parents went abroad for additional educational resources. The children hoped to follow parents abroad if they performed well at school. Remittances outweighed the detrimental effects of living in a migrant family (Botzat and Pfeiffer 2014). However, indirect effects of migration were acknowledged. These related to health and psychological depression depending on the structure of the family of the caregiver and the extent of involvement of the extended family in complementing the gaps resulting from parental migration.

Dietz et al. (2015) discovered that the remittant depended economy in Tajikistan channelled 1.8 billion dollars to education. About 94% migrants' children attended school, 63% in households had secondary education, 17% with tertiary and 16% had basic education. Every 7th household received remittances from abroad. The biggest funder of education were remittances. Inferior outcomes for education of children left behind had largely to do with less supervision by adults and the burden around the home. Among 7–11 years, 12–17 years school attendance was not consistent; otherwise, remittances had positive impact on education.

Bucheli et al. (2018) also found that in Ecuador, education accounts for 18% remittances. Enrolment rates were close to 100% before 11 years and were in line with mandatory years of schooling.

Despite notable positive effects of migration mentioned above, the opposing school of thought holds that migration has social costs resulting from a vacuum left by migrant parents. There are contradictions between material benefits and the social costs relating to socio-psychological and educational effects of migration (Baker et al. 2009). Women's federation discovered that left-behind children performed badly at school in the Chinese city of Pinaxing. Less than 20% achieved distinctions and 80% were on average or below. Some got 30–40% on compulsory courses. There was low attendance and decline in performance after the migration of parents. Those staying with elderly parents showed worst drop in marks (Wei Lu 2011).

Rural China witnessed changing roles of children resulting from change of the family structure because of migration (Breigenzer 2014). Children took up filial duties of taking care of the elderly at an early age because of absence of parents due to labour migration. Migrants' children viewed their migrant parents as strangers, resented them and felt unloved. About 23% of children in rural China dropped out of school before the end of grade 9, China's compulsory education level. The drop was a result of child labour to supplement incomes. A child who drops out of school early is likely to repeat the same as a parent forcing the children to be caught up in a cycle of dropping out of school.

In Moldova, children quitted schooling, lacked motivation for learning and about 22% absconded lessons. Academic performance deteriorated due to vulnerability to drugs, peer pressure and other illegal acts because of lack of supervision by parents (UNICEF 2008). In Sri Lanka, assessment registers showed that 28% of migrants' children did not attend school regularly and their educational achievement was lower than that of non-migrants' parents. About 75% of migrants' children were in the lowest bracket and all of them needed help in and outside the classroom (Perera and Rathnayaka 2013).

In some migratory households, the level of education is lowered by reduced aspirations for higher education. The possibility of moving abroad to find a job that does not require educational credentials hence discourages progress in education in Lima and Peru (Edward and Ureta 2003). The Latin American Migration Programme (LAMP) also found that the hyper-inflation and unemployment as a result of structural reforms in Peru and Lima intensified international migration causing low quality of primary and secondary education (Oropesa and Robles 2011). Although enrolments and retention increased among 15–20 year olds in migratory households, advancement was negatively affected. The children of migrants who were above 18 years desired to migrate because of motivation from positive returns of migration.

From 2000 to 2007, remittances increased from $700,000 to over 2 billion dollars causing high migrant flows. Besides remittances, there are other ways in which migration in Peru and Lima affected schooling. Where remittances were insufficient, children became potential workers. Left-behind children faced pressure within the family to leave school and enter the labour market. Migration of a family member reduced aspirations for higher education and expected returns in education, thereby fostering a family culture in which migration is viewed as an alternative economic mobility. Directly or indirectly, children are discouraged in their educational progress because of the possibility of moving abroad to find a job which neither requires nor rewards educational credentials. Other negatives include lack of monitoring of school attendance, performance and participation in the curriculum as well as the propensity of children to consume drugs (Oropesa and Robles 2011). Botezat (2018) concur that investment in material resources maybe better off but deprivation of primary caregivers was harmful to the well-being of left-behind children in Eastern Europe, Asia and South America. Absence of parents was associated with emotional and psychological distress.

Remittances have become a source of income in developing countries including Zimbabwe. As a result, there has been growing focus on quantity, benefits and use of remittances, thereby making international cross-border migration favourable (Dodson et al. 2008). Remittances can be defined as household income from foreign countries arising mainly from temporary or permanent movement of people to those economies (International Monetary Fund in Banya 2018). Developing countries receive an estimate of US$80 billion annual remittances and the figure can be higher because some are channelled through informal means. Remittances are viewed as an important source of finance for many African households (Maphosa 2007). In 2016, migrant remittances in developing countries accounted to US$429 billion. The flows into Sub-Saharan Africa stood at US$34 billion in 2017.

Zimbabwean population abroad is credited for helping the country's economy through remittances. In 2015, remittances for Zimbabwe was US$935 million. Zimbabwe ranks high in global remittances. In 2016, it ranked 36th in the world ranking and 5th in Africa after Nigeria, Egypt, Morocco and Ghana (World Bank in Banya 2018). The total remittance for Zimbabwe in 2016 was US$1853 billion, of course including humanitarian assistance. It must be noted, however, that the statistical data is not very accurate because some remittances come through unofficial means. Most of the illegal migrants use informal money transfers. They themselves and their informal transfers are not documented (Banya 2018).

Migration and Remittance Survey (MARS) noted that remittances have a hidden economic value of labour migration in Zimbabwe. Remittances ease household budget constraints and enable people to meet immediate needs like food, clothing, housing, health care, transport and agricultural implements. Forty-eight percent of remittances are used to pay school fees after immediate needs have been met (Tevera and Chikanda 2009). Remittances sent formally and informally are important for the development of infrastructure in the source country.

In Tsholotsho district in Matabeleland North in Zimbabwe, remittances are a main source of income of more than half of the households and they reshape the local economy. The remittances are used in agricultural farming equipment, livestock, dealer and bottle stores, grinding mills as well as scotch carts, among others. There is also investment in modern technology like solar lights to charge radios, cell phones and refrigerators. Three-quarters of households have at least a member in South Africa and, on average, receive (ZAR500) South African rand per month. Remittances benefit local development infrastructure and businesses. Very little is spent on social services like education and health. As a result, in Tsholotsho, there is no strong association between remittances and schooling (Ncube and Gomez 2011).

Ndlovu and Tigere (2018) discovered that in Lower Gweru, Zimbabwe, migrants found it difficult to send sufficient remittances back home. Families of migrants continued in poverty. Migrants could exceed a year without coming back, and life had not changed much for migrant families because remittances were very little. Children of migrants did not pay school fees and development levies on time. They were also not provided with proper and complete uniforms and other educational materials, money sufficed for basic foodstuffs. Sometimes relatives diverted it for personal use.

Children lacked parental care resulting of behaviour problems associated with lack of supervision and control at home. Those who misbehaved had trouble with authorities at school. They were rowdy, disrespectful, stubborn and unwilling to learn. Although incidences of dropping out are not high, absenteeism and poor performance is high. Lack of assistance at home, being left with illiterate elderly caregivers compromise educational outcomes.

In contrast to the findings by Maphosa (2007), Ndlovu and Tigere (2018) as well as Ncube and Gomez (2011) that purport that very little remittances are spent on social services like education in Zimbabwe, qualitative study findings in Bulilima and Mangwe as elsewhere, reveal that remittances increase enrolments and retention. The results are not quantified but sentiments from children, teachers and caregivers testify to this fact. Remittances had positive effects that relating to enrolments, retention and provision of educational materials. However, it is absence of parents, lack of children supervision and psychological stress that impacted negatively on the education of left-behind children.

Positive Effects of Migration in Zimbabwe: Bulilima and Mangwe

South Africa transfers a significant value of remittances to the Southern districts of Zimbabwe, Bulilima and Mangwe. Remittances, to a greater extent, are not used for educational fees in Matabeleland South. The largest proportion is spent on consumption, for example, clothing, food, electrical goods, scotch carts, brightly painted luxury houses, improvement of roads and bridges (Maphosa 2007). However, of late, migration to especially South Africa and Botswana has positive effects on educational enrolment, retention and provision of educational materials. There is a strong relationship between migratory remittances and schooling through increased enrolments and retention. If it were not for migration of parents, there would be no school to talk about taking into account the geo-physical and socio-economic marginality of the area.

Generally, remittances increase asset accumulation, home ownership, health, nutrition, income disposal and access to education. However, the development impact is felt more at household level than the community. Remittances from migrants in South Africa have improved the quality of education for children in Zimbabwe. At least 51% of migrants can afford extra money for tuition, extra textbooks and other teaching and learning equipment used by primary and secondary school children. Some are able to send children to private schools where their children get better education and make choices to write Cambridge, Zimbabwe School Examinations Council Examinations (ZIMSEC) or both. Cambridge is preferred for its international recognition. Migrants' children have better resources for education (Moyo 2016).

Migration is the sole source of income for schooling and livelihoods for families. Fees, stationery, food for upkeep and uniforms for schooling are all from migratory remittances. Without migration, migrant parents would not be employed. They would not afford to pay fees and to support their families. Labour migration in Bulilima and Mangwe eases household budget constraints.

South West Zimbabwe is a poor region with no sources of income and with escalating unemployment. Migration remains the only option. The uniforms, stationary are easily accessible and cheaper outside Zimbabwe. Most of those who afford to send children to school are migrants. In fact, three-quarters of learners in mission, government boarding and low-cost boarding have parents who are migrants. Even the status and conditions of life of migrants and their children is far much better than those of non-migrants. Most of the fees collected for school development funds are from migrant parents. Migration is beneficial in terms of educational provisions.

Migrants pull resources together to improve school infrastructure, classroom, blocks, toilets, ablution facilities and dormitories for learners in low-cost boarding schools and the general running of the low-cost boarding schools. Matrons and boarding masters are paid by parents, most of which are migrants. Migrants have played an important role in transforming former "bush boarding" into low cost of course with other non-migrants' parents. It must be noted, however, that the majority of learners in these low-cost boardings are migrants' children.

Negative Effects of Migration

Ndlovu and Tigere (2018) discovered that life did not change much for left-behind families despite remittances in Lower Gweru, ward 8, Zimbabwe. About 11,620 migrants left the country for the diaspora as a result of soaring official rate of inflation reaching alarming levels of 231 million percent making life unbearable to many Zimbabweans. Having arrived in destination countries, migrants discover that it is difficult to send sufficient remittances for the people left behind. Sometimes they fail to pay back the money borrowed for travelling or to replace livestock sold for transport. Remittances are very little and families continue to sink in poverty. They exceed a year before coming back home mostly because there is nothing to bring home. Elderly people and their children go to clinics unaccompanied. Some children default their medicines because of lack of parents or guardians due to migration. Children do not pay full fees in time in schools. They are not provided for with school uniforms and other educational materials. Money from remittances manages to cover basic foodstuffs and sometimes relatives divert it to their personal use.

There are also challenges associated with behaviour patterns and lack of parental supervision and control that tend to compromise the benefits of migration. Children engage in alcohol and delinquent activities. They are rowdy, stubborn, very disrespectful and unwilling to learn. In most cases, they are in trouble with school authorities. They do not obey elderly parents and caregivers whom they stay with. Although they do not drop from school, there are high incidences of absenteeism and poor performance. Poor performance relates to lack of discipline and educational assistance at home. Lack of assistance by illiterate caregivers compromises educational outcomes. Generally, international migration has undesirable effects on left-behind children in ward 8, Lower Gweru, Zimbabwe (Ndlovu and Tigere 2018).

In Bulilima and Mangwe, as in other parts of the country not all families benefit from migration. To migrant parents who either abandon families or are unable to support families, migration has no benefits. Some migrant parents or guardians are *imigewu* or *imitshoni. Imigewu* is derived from the Ndebele word *umgewu* meaning a person who travels for good and never returns. *Mtshoni* is a noun derived from the verb *ukutshona* meaning going for good or disappearing. Some migrants abandon families and die as migrants while some return in old age after having wasted profitable years and energy. Their migrant families live in abject poverty and fail to access schooling.

Factors such as unreliable jobs in the destination country, distance from home, family expenses, delays and in transportation of money and goods by *omalayitsha* as well as illegality of stay of migrants in destination countries make it difficult for some families to benefit from migration. Omalayitshas also charge exorbitant prices for transporting goods. Some families suffer and benefit nothing from migration. Not all children and families of migrants benefit from migration.

The changing economic situation of the rand currency and of late, the devaluation of the rand and the local bond note versus the US Dollar make many families not to afford the fees resulting in learners dropping out of school. Not all migrants can afford fees. Some migrants' children are in and out of school frequently and this situation is stressful. The illegal means used by parents to transport remittances are unreliable causing delays in payment of fees, arrival of educational material and uniforms.

Although migrants' children with responsible parents benefit economically in terms of support, food, clothing and general provision for livelihoods including fees and educational materials despite delays, the negative effects of migration are greater than the material benefits. Lack of parental guidance and motivation of migrants' children is detrimental to children's education. Some children are left in child-headed families or with employed workers or elderly people who cannot assist or give guidance on educational issues. The education of children demands that parents and teachers work together. The dual social support in guidance, care and love is vital in the education of children.

Migration of parents burden children with work. Migrants' children bother teachers on social issues of child development, care issues and problems they have with caregivers at home. Learners who are migrants' children now view teachers as parents. Issues that are supposed to be reported to parents are reported to teachers.

Migrant parents are well aware of the fact that leaving children behind is a problem, but they cannot travel with them because they themselves have no work permits and therefore cannot obtain study permits for their children. Parents phone teachers asking them to assist in counselling children who are in mission or are in low-cost boarding on issues like budgeting and exchange of currencies. Children need assistance in this area, but have no one ready to assist. Parents also ask teachers to buy stationery and groceries for their children over the weekends fearing that if they give such responsibility to relatives, money may be used on other items that have nothing to do with children's schooling. By migration, parents are like shooting themselves in the foot as far as education for their children is concerned. It is exchanging education and social development of their children for improved lifestyles and material benefits.

Caregivers marginalise and neglect the roles of parental guidance, supervision, and security. Learners commuting from their rural homes and those lodging in townships are prone to abuse, have too much freedom which in turn affects performance and completion of schooling. The material benefits of migration are at the expense of character building. Children are vulnerable and have no one to talk to when they face complex situations of vulnerability.

Caregivers are burdened by children of migrants because not every migrant remits financial and educational resources in a satisfactory manner. There are three categories of migrant parents; a few who remit satisfactorily, the struggling and the irresponsible who do not remit. However, some parents make concerted efforts to provide everything needed particularly for children in mission and low-cost boarding. They take up the responsibility and send everything in time.

Others are committed to sending remittances, but these are insufficient because of the types of jobs they do in destination countries. Sometimes they send once per term, once in two months or monthly but very few items. They struggle to provide support for two homes (diasporan and Zimbabwean, school fees and examination fees. Sometimes children fail to proceed to lower sixth, college or university. Others seem to be struggling for survival in destination countries, therefore unable to cater for all the needs for their children. They end up borrowing money from caregivers some of which they sometimes fail to return.

Effects of Migration on Parents and Learners' Attitudes Towards Migration

Migration increases aspirations for migration for some of the migrants' children. It is the only option in Bulilima and Mangwe's marginalised districts where there are no alternative economic activities. Some of the children wish to follow their parents after completing Ordinary and Advanced Levels because they feel they are well catered for by parents and relatives who are outside Zimbabwe. However, not all learners wish to go but only those who have been influenced by the culture of migration. Those who see value in migration drop out of school to migrate. Learners' conversations and their language indicate envy and desire to migrate. They communicate with friends and teachers in language of migrants, *isinjiva. Isinjiva* means a language derived from the term *njiva,* a common term used in the country to refer to migrants. Learners boast that in no time they would be seen driving *iviri, imoto or a Gauteng* meaning a car from the Gauteng province.

Some want to migrate with parents and desire that there be a policy suggesting that parents should migrate with their children to further their schooling in the diaspora especially South Africa. These learners think South Africa is a paradise compared to Zimbabwe where there is hunger and unemployment. Parents are hailed for having migrated and they feel that if parents had not migrated, they would not be attending school and life would be difficult even for other social needs besides fees and educational materials.

Some male children do not see the need of having parents around. To them, what is important are material benefits especially those left with mothers and grandparents. Mothers and grandmothers are preferred as caregivers because they are less strict than male figures. Those in boarding and low-cost boarding also did not mind absence of parents because life in these environments is comfortable.

The years 2014 and 2015 saw a drop in learners' migration to South Africa and Botswana owing to the political and economic situation in South Africa. The fall of the rand currency, xenophobia attacks and expiry of asylum documents, lack of accommodation, illegality of stay, bad living conditions, violence and the general lifestyle caused decline in migration. Most of the learners who had stayed with parents in South Africa, whose asylum papers had not been renewed were brought home and are now in schools in Bulilima and Mangwe.

Most of those who have stayed or visited South Africa are not in favour of migration and are more serious with their schooling. It must also be noted that those who strongly oppose migration are learners in boarding, low-cost boarding and those from growth points and business centres. These learners have exposure and interact with educated professionals, for example, teachers, nurses, police officers, agricultural-extension officers and district development officers among others. This accounts for different attitudes.

A few who desire to migrate wish to do so after schooling owing to high rates of unemployment in Zimbabwe; otherwise, they are reluctant to migrate. Learners prefer working in than outside Zimbabwe if employment is available. Those who have no hope of passing or getting a job in Zimbabwe harbour the thoughts of migrating. Migration is an expectation to those who have a low priority given to education.

However, prevalence of dropouts increases owing to failure to pay fees by parents as a result of the fall of the rand currency versus the US Dollar. Most of the parents who fail to pay fees especially in mission and low-cost boarding are those in Botswana and South Africa. The dropouts increased towards the end of 2015 and early 2016. In particular, parents were failing to pay examination fees of up to 8 or 10 subjects at $10-00 for Ordinary Level and $15-00 for Advanced Level examinations, amounting to $100-00 or $150-00 to be paid simultaneously. The year 2018 to 2019 has also seen a rise in dropouts owing to political and economic situation in Zimbabwe. The rise in examination fees is likely to increase the dropout rate. Very few parents can pay $90 per subject for 6–10 subjects.

Boys who drop out as a result of unpaid fees and examinations, act as caretakers in migrant homes or herd cattle for neighbours. A few cross the border to Botswana because the pula is better than the rand. Girls who drop out of school as a result of failure to pay fees and examinations, stay at home or get married to njivas. Those from child-headed families drop out mostly because of pregnancy. Dropping out of school has to do with the economic circumstances more than the desire to migrate. Dropping out of school has resumed again in 2019–2020 because of the change of government and devaluation of the bond note versus the US Dollar. In the current years, migration by youths and the middle-aged people in this region has more to do with economic trends than the culture of migration as earlier highlight by Maphosa and Marojele (2013).

Effects on Attitudes Towards Schooling

Learners wish to advance to tertiary institutions and vow that they can go outside the country only when education and attempts to get jobs in Zimbabwe have failed. It is poverty and failure to pay fees by parents that force them to drop out of school to migrate. Otherwise, they desire to go as far as money and intelligence can take them. Some are repeating ordinary levels, evidence of commitment to education. With regard to career aspirations, migrants' children value education and look forward to be nurses, police officers, medical doctors, lawyers, pastors, engineers, journalists, mechanics, pilots, economists, accountants and economists among others.

Teachers, however, think that migrants' children have a negative attitude towards schooling. They feel lack of guidance affects concentration levels by learners, performance and consequently advancement with schooling. Teachers think most children are satisfied with passing ordinary level. They do not see value in educational advancement despite the fact that some of them are well provided for. Those who once migrated with parents are more negative than those who started schooling in Zimbabwe. Those who once migrated, look down upon teachers and view them as poor.

Of interest to note is that learners, teachers and caregivers collude that in Bulilima, Mangwe, migration has positive benefits as far as livelihood, and educational resources are concerned. However, positive effects are overshadowed by the negative impact of lack of parental guidance. Quality of educational outcomes go beyond increased enrolments, retention and provision of educational resources.

Migration is not an option for uneducated parents but the only source of livelihood and schooling. As a result, children do not see beyond the material benefits. Some children do not realise that not all migrant families benefit from migration but only a few with remitting migrants. Children of non-remitting or struggling migrants do not benefit to a great extent. Although children support parental migration, they are against parents who migrate with children of school-going age. They prefer to complete schooling first. Migrant parents and their children have positive attitude towards education.

Conclusion

The problem is that increased access and retention to schooling as a result of remittances in Zimbabwe is overshadowed by lack of parental guidance which affects migrants' children negatively. The increased benefits in terms of remittances for schooling are at the expense of quality education, performance, child development, protection and security. At school, there is a strong relationship between remittances, increased enrolments and retention in schooling. Schools are populated by migrants' children. The benefits include fees and other educational resources. In fact, the conclusion is that without migration of parents there would be very little to talk about in terms of schooling. At family and community level, migration is the sole source of livelihood and improved lifestyles, including sending children to school. Remittances from labour migration by parents expand the budget to include fees, stationery and other educational resources. Parents try to compensate for their absence by fostering children to extended family caregivers, provision of educational materials, presents to motivate their children and to ensure that children's needs are taken care of.

References

Baker C et al (2009) The impact of migration on children in the Caribbean. Available from http://www.unicef.org

Banya N (2018) Do Zimbabweans abroad send US$1 billion home annually? ZIMFACT. Available from https://zimfact.org. Accessed 20 Apr 2020

Botzat A, Pfeiffer F (2014) The impact of parents' migration on the well-being of children left behind: initial evidence from Romania. Discussion Paper no 6225. Mannheim: Centre for European Economic Research.

Botezat A (2018) Parental migration and children left behind in Eastern Europe, Asia and South America. Bold https://bold.expert/parental-migration and the children-left-behind

Breigenzer A (2014) The effects of labour migration on the traditional Chinese family structure and the role of children. Thesis, Washington University. https://research.libraries.wsu.edu

Bucheli et al (2018) Mixed effects remittances on child education. IZA journal of development and migration. https://linkspringer.com

Diet et al (2015) Emigration, remittances and education of children staying behind: Evidence from Tajikistan. Discussion paper no 9715. Born: IZA

Dodson BJ et al (2008) Gender, migration and remittances in South Africa. SAMP, Cape Town

Edward A, Ureta M (2003) International migration, remittances and schooling: evidence from El-Salvador. NBER Working Paper 9776. Available from http://www.eduglossary.org. Accessed 10 June 2013

Federici D, Gannett M (2006) Temporary migration and foreign direct investment. Springer Science and Business Media, Rome

Maphosa F (2004) The impact of remittances from Zimbabweans working in South Africa and rural livelihoods in the Southern districts of Zimbabwe. Available from http://codesria.org. Accessed 29 Dec 2013

Maphosa F (2007) Remittances and development: the effects of migration to South Africa and rural livelihoods in Southern Zimbabwe. Routledge, London

Maphosa F, Marojele, Marojele RN (2013). Changing the culture of migration: A attitudes towards education among the Basutho labour migrants to South African mines. African Development, XXXVIII (1 and 2), 151–170

Moyo I (2016) Remittances and development: Zimbabwe migrants' teachers in South Africa and their impact on the Zimbabwean families. Available from https://ups2015.princeton.edu. Accessed 20 Apr 2020

Ncube G, Gomez GM (2011) Zimbabwe local economic development and migrant remittances in rural Zimbabwe, Tsholotsho District. International Institute of Social Studies. Working Paper no 523. Available from http://www.repub.eur.nl./pub. Accessed 2 Feb 2014

Ncube, G and Gomez, GM (2011) Local economic development and migrant remittances in rural Zimbabwe. URL ISS Working paper series/general series, 2011-repub.eur.nl

Ndlovu E, Tigere R (2018). Economic migration and socio-economic impacts on the emigrants' family: a case of ward 8 Gweru rural district, Zimbabwe. Jamba J Disaster Risk Stud 10 (1). Available from https://oloi.org/jamba.voio/414. Accessed 20 Apr 2020

Oropesa RS, Robles FV (2011) International migration and the children: evidence from Lima and Peru. Available from http://www.springerlink.com. Accessed 8 Apr 2012

Perera V, Rathnayaka MR (2013) Sri Lanka's missing mothers. A working paper on the effects of mother migration on children. Save the Children, Baticoloa

Rossi A (2008) The impact of migration on children on developing countries. Centre for Human Resources Policy, Cambridge

Tevera D, Chikanda A (2009) Migrant remittances and household survival in Zimbabwe. Migration Policy Series. No. 51. IDASA, Capetown.

UNICEF (2008) The effects of parental deprivation on the development of the child left behind by Moldovan migrants. UNDP/CRIC, New York

Wei Lu G (2011) Left-behind children in rural China: research based on the use of qualitative methods in inner Mongolia. University of New York, New York

Chapter 5
Educational Challenges of Migrants' Children at Home and School: Teachers, Learners and Caregivers' Perspectives

Cultural Capital and Cultural Deprivation Theory

The cultural capital and cultural deprivation theories illuminate educational challenges of migrants' children at home and school. Cultural capital determines their educational outcomes and explains ways in which relations of marginality and inequality are reproduced in the family and the education system. Cultural capital is composed of cultural habits and dispositions inherited from the family and transmitted from one generation to the other (Boudieu in Laureau and Weininger 2003). These include elaborate linguistic codes of the dominant class, attitudes, practices, beliefs, skills, aspirations and motivation (Beinstern 2001).

Cultural capital can be classified into four categories as follows:

- economic or material capital which enables one to acquire educational resources
- social networks in the education system or workplace
- Institutionalised cultural capital gained through the family environment with advantages of education, for example, coming from families with educated people
- Symbolic capital which involves knowing the right persons (Bourdieu and Paserron 1997 cited in Giddens 2001, p. 46).

Migrants' children lack all the types of cultural capital. Rural migrants' children are vulnerable because they do not belong to the dominant class. Most of their parents and caregivers are poor and illiterate and have no educational connections with the dominant class while schools as institutions and teachers expect these children to have educational values and attitudes of the dominant class. Absence of parents and illiteracy of left-behind caregivers does not create environment for favourable educational outcomes.

The education system advantages those who have cultural capital from their home environments. Poor lower class migrants' children with absent parents have a culture of disadvantages, deprivation, under-privileged or culturally induced

W. Matsa, *Marginality, Migration and Education*,
https://doi.org/10.1007/978-3-030-60873-6_5

backwardness. Cultural deprivation is the lack of "material and non-material (cultural capital) which affects the individual's ability to communicate perform and respond appropriately in formal institutions" (McCrossman 1966, p. 6).

Lack of cultural capital in the home often translates into lack of educational stimulation which is key to children's education. Cultural capital influences desirable habits of schooling such as reading preferences, parents' participation in the education of the child, aspirations for educational advancement as well as career aspirations. In this book, the impact of cultural capital and cultural deprivation are explored to unearth their effects on migrants' children in Bulilima and Mangwe, Zimbabwe.

The funds of knowledge explain learners' own adaptation strategies in the face of challenges at home and school because of absence of parents due to migration. Despite cultural deprivation experiences, migrants' children in Bulilima and Mangwe developed funds of knowledge that enabled them to adapt and survive in such living and educational environments. Funds of knowledge refer to the essential cultural practices and bodies of knowledge and information that households use to survive, get ahead or thrive (Moll and Greenberg 1990, p. 32). Funds of knowledge approach alter the perception of the marginalised children, families or communities. Funds of knowledge enable people who are in challenges to use knowledge emerging from their vulnerable life experiences to cope with challenges. Migrants' children developed a sense of responsibility and coping strategies for survival.

Educational Challenges of Migrants' Children Elsewhere

Migrants' children experience challenges ranging from scarcity of resources like food, educational materials, health medication, shelter and financial support because remittances are sometimes insignificant (Cortes 2007; Gassmann et al. 2003; UNICEF 2008; Xiang 2005). Migrants' children also experience challenges of deprivation of supporting school and home environments, educational assistance by caregivers and educational supervision at home (Ganju 2012; Perera and Rathnayaka 2013). Other direct challenges, according to Tufis (2007), in Romania include lack of role models in the area of education because most of the children do not even know what their parents are doing abroad therefore are not inspired to perform well academically. Children of non-migrants are likely to take on their parent's professions. Migrants' children also lack mentors or supervisors to pay attention on time spent on homework, study time, whereabouts, attendance, punctuality, attitudes and interest as these largely depend on the home situation. They also lack link persons between them and teachers at school. The problems of relations with caregivers, material and financial resources may not be known at school.

In North and south Moldova, migrants' children encounter deprivations and repercussions of separation with parents. Psychological problems that are noted include worries, loss of interest in school, loneliness and lack of attachment. Their life is characterised by longing, sadness, loneliness and feelings of insecurity because caregivers are not close enough (UNICEF 2008; Gavriliuc 2006).

In Georgia, absence of mothers resulted in loneliness and some children had emotional distances with their fathers, girls were better off than boys. Children of migrants looked frail. The negative effects of migration are larger than the beneficial effects of remittances (Luecke and Stoehr 2014). Parents send luxurious goods like radios, televisions, solar batteries and clothes rather than educational materials. Children also lack medical assistance and reproductive health knowledge, thereby making them vulnerable to health risks that end up affecting schooling.

In the two districts of Indramayu, villages of Juntikebon and Juntikedokan senior citizens of West Java had challenges with migrants' children. They had health problems associated with old age and unable to do all the duties related to nurturing. About 59% were above 60–69 years, 30% ranged from 70 to 79 years while the rest 9.3% were above 80 years. The absence of parents generated deterioration of health conditions. The senior citizens did not have and could not prepare nutritious food for the children. They did not have health insurance schemes or social safety nets. Migrating parents did not put senior citizens and their children on health insurance schemes. They also did not seek consent of senior citizens on care of their children but assumed that caring was their natural role (Noveria 2014).

In the Caribbean, children experienced bad moods, isolation, loss of appetite, anxiety, low concentration, tantrums and peer pressure (Save the Children 2014). Tufis (2007) echoed interpersonal sensitivity, hostility and deviant behaviour, feelings of sadness, difficulties in concentration and isolation.

Children living in poor households in Sri Lanka had chances of completing junior school lowered. Boys have a high probability of migrating because they have to migrate to sustain families in these poor households. As a result of socialisation, labour force demands are placed on the boy-child at an early age. Although girls remain at school, they also have household duties. In the rural Chinese provinces of Hubei, Anhoux, Heran and Sichuan the majority of elderly caregivers could not tutor learners because they were either uneducated or the education they had was different from that of their grandchildren. Some cared less about education, very few were willing to tutor migrants' children at home, even if they could. This was because the nature of responsibility was not clarified to caregivers especially the issues to do with educational supervision and discipline as well as psychological well-being of children. Caregivers, especially grandparents had no push for children to attend and complete school. Some underestimated their important influence on children's education. They thought education was the responsibility of the schools; theirs was the upkeep at home; shelter, food and security. Some children in the four provinces held the belief that school was useless, a belief held by some caregivers in the area. The values of the community and caregivers influenced children's attitudes towards schooling.

Children also lacked exposure because caregivers did not take children out to broaden their horizon. There is need to widen their knowledge beyond villages in which they live. The majority of children had narrow social view of the world because they were forced to stay home for security reasons. Some caregivers were too strict with children to avoid risks of car accidents and abductions. Children were confined to the homes most of the times. Those who stayed with caregivers who were not strict developed bad habits that made them vulnerable to criminal activities. Those who lacked self-discipline in Sichuan province did not attend school, were extravagant and experimented on drugs making themselves vulnerable to various injuries.

Zimbabwe National Strategic Plan (2010) discovered that in Zimbabwe scarcity of resources was a formidable educational challenge to migrants' children. In extreme economic marginality, orphans and migrants' children were withdrawn from school to work in the fields for survival. Girls are employed in restaurants, nightclubs and grocery stores where there is high risk of abuse, HIV and AIDS as well as exploitation. Some girls are employed as domestic workers or tricked into accepting dubious employment arrangements and become vulnerable. Some girls engage in child prostitution with truck drivers along Harare-Beitbridge, Harare-Chirundu and Harare-Mudzi roads. In Mt. Darwin, Mashonaland Central of Zimbabwe, more boys drop out of school to engage in cross-border trading in bordering countries or become illegal cotton or tobacco pickers, graders, fruit collectors as well as gold panners. In Tsholotsho, Matabeleland North which shares border with Bulilima and Mangwe, harsh economic conditions force boys to follow the exodus to South Africa and Botswana leaving school. As a result of scarcity of resources, migrants' children are forced to take the role of breadwinners at the expense of schooling.

Educational Challenges Experienced at Home in Bulilima and Mangwe, Zimbabwe

The major challenges identified were delayed fees, lack of time for studies, shortages of educational materials, ill-treatment by caregivers, lack of moral and practical educational assistance at home as well as lack of parental guidance and control which are key in children's education. Lack of educational assistance and ill-treatment by caregivers are major challenges at home. Basic needs and food shortages as well as communication are also challenges experienced by migrants' children in their homes.

Most children have nasty experiences at home because not all migrants have good jobs. Some are neglected children who live alone or with uncaring relatives and employed caregivers with no solutions to their problems. Those who stay with their biological mothers or able-bodied grandparents have fewer challenges. Mothers and grandparents take care of the challenges.

Children in child-headed families make decisions concerning all spheres of life alone. These include academic life, for example, when to wake up for school, whether to go to school or not, payment of fees, budget, educational resources and homework among others. Some have all the resources, quality stationery and uniforms but lack parental support in decision-making like arranging for extra lessons, parental moral support during consultation and prize-giving days.

Lack of study time is a result of the burden of household chores and distances travelled to and from schools. Some migrant children leave for school at dawn and arrive back home in the evening. Some caregivers deliberately deny them study time arguing that because they spend the whole day at school, they should not expect caregivers to do everything for them. Most of those who stay alone are burdened with household chores to the extent of coming to school with unwashed uniforms or are frequently absent from school.

Shortage of educational materials is a result of delays in fees, transportation and delivery of goods especially from South Africa and Botswana. The delays affect food supplies and other basic needs that end up affecting schooling directly. Food shortages trigger ill-treatment by caregivers who then insult migrants' children for poor budgeting. Sometimes they tell migrants children that their parents are not sending frequently and are not sending enough. Caregivers vent words implying that taking care of migrants' children is burdensome. In such cases, caregivers then make no efforts for the provisions of morning meals or packed lunches, forcing these children to spend the whole day on empty stomachs at school. The shortages also have ripple effects to absenteeism, anxiety, stress and other challenges such as lighting systems (candles and paraffin), as well as lack of care and love, all of which affect schooling directly or indirectly. There is generally emptiness, sadness, irritability, resentments, rejection moods, silence, denial and displacement of activities among children of migrants in Zimbabwe as observed by (Filippa 2011).

The above "lacks" are necessities for quality education. The lack of these important necessities is cultural deprivation (McCrossman 1966, p. 6), yet to Beinstern (2001, p. 2), access to these constitutes cultural capital for schooling which determines educational outcomes. Migrants' children lack economic cultural capital which enables children to acquire educational resources and institutionalised capital which has to do with the school or family educational supportive environments. Fees, basic needs and educational resources are all economic resources, the cultural capital to meet schools' expectations.

The "lacks" at home translate into deprivation at school thereby affecting motivation for schooling. Shortage of food, time for homework, educational assistance and guidance at home as well as lighting systems all are institutionalised capital deprivation of a conducive environment for schooling. Schools do not take these challenges into consideration but ignore such issues as marginal. Migrants' children and non-migrant children receive the same treatment at school yet non-migrants and migrants' children have different home and schooling environments.

Transportation delays also worsen food supplies, educational materials and in the community, it is embarrassing not to have a complete uniform and quality stationery, hard covered exercise books as well as food to carry to school if the parents are *njivas*. At home, if there are shortages of food, educational materials like pens, there is no one to report to. If children alert caregivers, it is inviting insults about their useless parents who send insufficient needs.

Delays in fees and other educational materials make learners experience serious challenges to which relatives or caregivers have no solutions. They simply phone the parents. In most cases, migrant parents do not get their December salaries because they would be on holiday. They get paid at the end of January and this result in late payment of fees. Migrants' children end up joining schools in early February after having paid their fees in full. Sometimes caregivers use money left for fees for important basic items like mealie-meal and wait for the migrant parents to send fees at the end of January. At school, administrators do not accept any excuses.

The two biggest challenges of migrants' children are lack of parental guidance and shortage of educational resources. The two give rise to other sub-problems, for example, violation of rights to care, protection, quality education, healthy and security. These create environments not favourable to schooling. Cultural deprivation at home led to lack of cultural capital for schooling.

Lacks of parental guidance, control, discipline and care have a bearing on children's schooling. Learners have no one to encourage them to study and have too much freedom, especially those in child-headed families. Lack of parental guidance results in over-spending which further worsen shortages. Children are also vulnerable to abuse and lack social security. They play the roles of adults doing work beyond their age, for example, managing households and taking care of younger siblings. Children staying alone are under-protected, inadequately supervised, exposed to harmful life and assume adult responsibilities prematurely.

Migrants' children are disadvantaged compared to non-migrant ones although teachers and caregivers played a blame game on the issue. Children's challenges are either hidden or denied by teachers and caregivers. Disadvantages at home are hidden from school authorities and teachers or they chose to ignore them. Caregivers concentrated on challenges at school and down played those at home. Teachers pinpointed challenges children experienced at home and trivialised those experienced at school.

Some grandparents, relatives and employed caregivers handle the migrants' children softly and do not report mischief early fearing that migrant parents might take the child and stop sending groceries. The comfort of residence and welfare depend on the relationship with caregivers. The learners who are better off are those staying with mothers, grandparents, brothers and sisters. Mothers and grandmothers are the most preferred for care and love. Sisters and brothers are the second best. Employed caretakers and other relatives, especially from the father's side seem cruel and uncaring. Some are jealous and report every minor mistake to migrant parents. Whenever they phone migrant parents, in most cases, it is about negative things done by migrants' children. Some relatives seem to have agreed to take care of the children in order to benefit from migrant remittances.

Learners think caregivers are remittance opportunists, economic swindlers who hide behind the notion of caregiving, when in fact; the aim is to get economic benefits from migrant parents. This is similar to what Kufakurinani (2014, p. 125) referred to as the "crisis of expectation" in Zimbabwe where caregivers expect remittances from migrant parents and migrant parents expect kin to reciprocate by looking after their children. Caregivers do not fill the caring gap in the home as expected by migrants' parents.

Illiteracy of caregivers among other things make it difficult for them to be link persons between the school and migrant parents or to honour important days of the school calendar-like consultation and prize-giving days. The elderly, especially grandparents desire to attend but illiteracy and distances to be travelled are barriers. Attendance by uneducated mothers and other relatives is not helpful because some cannot even read. Uncles and employed workers do not bother even if educated. Sometimes lack of attendance is a result of instructions from migrant parents, fearing gossip about their children by jealous relatives. Relatives and caregivers sometimes exaggerate or twist messages from teachers to paint bad pictures of learners with absent parents to the whole community. Sometimes neighbours attending consultations for their children are asked to consult on migrant parents' behalf, but these do not bother to check books or mark lists. This does not serve the purpose of consultations.

Migrant parents try all they can to give long distance support by encouraging their children to work hard. They phone to ask about performance, try to be quick to respond to school needs and motivate their children through incentives to pass. Incentives include increased pocket money, groceries, and clothing for Christmas and holiday visits to migrant parents' destination countries. Sometimes cell phones, iPads and wristwatches are bought for them. They also threaten to withdraw privileges if learners do not take education seriously.

All these efforts are marred by distance and absence of parents. Some of the learners are well financed and have all the material educational need, what is lacking is parental support and a favourable environment at home. Caregivers rarely talk about schooling and never look at their books. Moral support of parents is crucial in education. Learners also have no one to help them with homework because most caregivers are illiterate. They expect learners to do everything related to schooling at school. They only ensure that they go to school every day and to report to parents any misconduct related to schooling. However, literate mothers, brothers, sisters and some foster parents are of help.

Some children commute long distances to school and the school programmes end very late. The teachers expect their homework to be done as given. When school programmes end late afternoon, some children arrive home around 6–7 pm. They have no time for homework and there are no people to assist. Teachers punish learners who do not do homework. Some wake up as early as 3 am to get time to copy homework from others in the morning at school. However, learners brave these hardships, develop adaptive strategies and complete their secondary education under such difficult circumstances.

Some of those who stay near the school with caregivers who are elderly, uncaring caregivers or those who stay alone have an additional challenge of burdens of household chores. Caregivers wait for learners to come from school and do the chores. If girls wake up early are sleepy in lessons, they are suspected by teachers to be pregnant. However, those who stay with employed caregivers, mothers and the able-bodied grandparents have less work. They find work done and food ready. Other relatives wait for them to return from school to do the work. The burden of household chores, sometimes result not only in lack of time for study, but also late coming, absenteeism and absconding of lessons. Household work disturbs attendance, time spent on homework, punctuality and consequently attitudes and interest in schooling. However, a few are able to combine work with schooling and manage to do well at school.

Extra lessons are expensive and recently illegal and can only be done during vacations. There are no libraries even during vacations and the lighting system is also a problem. Migrant parents remove their solar systems or battery lights when returning after the holidays fearing for abuse by relatives. Instead of teachers trying to understand the situation of migrants' children, they add salt to injury by severely punishing children for coming late and for not doing homework. Teachers beat, insult or pass negative comments as a result. Children have developed ways of going around each problem. They help each other in homework, form study groups among themselves, do most of the chores during weekend and develop their own networks with teachers for cheaper or free extra lessons.

Challenges of Migrants' Children at School

The challenges of migration on migrants' children at school can be categorised as psychological (emotional), material, social and educational marginalisation. Their basic and educational, care and protection needs are treated as marginal issues by parents, teachers, caregivers and the community at large thereby marginalising them. Migrants' children have challenges at school that range from late payment of fees, lack of study time at school, lack of educational support and educational material for schooling.

Delayed fees affect both the start of the term and access to educational materials. However, shortages are particularly serious towards the end of the term. This period is characterised by shortage of stationery, going to school on empty stomach and for boarders, shortage of toiletries and reliance on the dining halls as the only sole source of food. Some learners run short of pens and pocket money to buy school necessities. Theft of quality stationary brought from outside is another cause of shortage of stationery. Both migrants and non-migrants' children steal from each other. Teachers also lose learners' new books through theft but expect learners to replace them. This is common at the beginning and towards the end of the term.

For migrants' children, this is a challenge because most of their stationery is bought in South Africa and Botswana. With the low exchange rate of the rand, they are finding it difficult to use their pocket money to replace anything. They have to phone their migrant parents and supplies are delayed by transportation through *omalayitsha*.

Learners fail to meet school expectations in terms of financial and material needs resulting in psychological problems and low readiness for schooling. They pay fees and levies in instalments. Parents take a long time to source and send educational materials and other needs. This coupled with over-spending and lack of proper budgets by learners and caregivers, leads to dropout of potentially good students. The lack of communication with parents further delays the sourcing and arrival of educational remittances. Late payment of fees and lack of other educational materials cause children's lives at school to be characterised by deprivation of cultural capital. Deprivations affect access to, retention in school and the quality of education thereby violating the right to quality education despite international calls for EFA (UNESCO 2012).

Absence of parents and lack of parental discipline make it difficult for children to take instructions from anyone because they are used to being their own bosses at home. They are seen as disrespectful to teachers and other children. They are also viewed as stubborn, rude and violent because they are not used to any rules. However, not all of the learners are bad. There are some who are well mannered.

Emotional and psychological moods indirectly affect schooling. Learners experience mood swing. At times, they are happy and content but at times, they appear stressed, tense, intolerable, and quick to be angry showing hostility towards teachers and other children and reluctance to take advice. Teachers and administrators do not bother to understand the situation of migrants' children. They send them home for incomplete fees, uniforms and shortage of stationery knowing that these would arrive by the end of the month. They do not accept negotiations even when their parents phone the school. Teachers chase them out of lessons for shortage of educational materials or incomplete homework or for coming late. They never take into consideration the distances travelled from home, lack of study time and lack of educational assistance at home. They treat rural day commuters as boarders and low-cost boarders who have ample time for study and assistance with homework. Parental deprivation leads to various forms of other deprivations.

Important days of the school calendar are not honoured by both migrant parents and caregivers. Elderly parents do not attend consultations, visiting and prize-giving days because of distance to schools. Able-bodied caregivers do not bother because they feel it is not their duty and some do not value education. Most migrant parents never attend because of distance. They send workers to consultation, prize-giving days and annual general meetings who sometimes do not respond. Some parents and caregivers marginalise invitations by the school authorities and never make attempts to send anyone. Some learners are visited once or never up to Upper 6 grade. When learners bring their brothers and sisters who are school leavers for consultations, teachers do not take them seriously. Success stories or problems of migrants' children are not known to their parents. They lack recognition and praises even for good things they do at school. Parents depend on hearsay for their children's performance and well-being.

Teachers and administrators view non-attendance of parents and caregivers to important functions in the school as lack of seriousness about education. Learners get stressed knowing that they would be beaten up or punished for non-attendance by parents. Sometimes they are asked to bring parents or to go and collect them in person. This is a challenge for those in child-headed families and those staying with the elderly or the uncaring. Most of what is viewed as cases of indiscipline at school or breaking of school rules are issues related to the situation of absent parents.

Lack of parental guidance and discipline have ripple effects, children develop certain behavioural characteristics that make it difficult for them to fit into the school system or to perform well. Some of the learners are rude, violent, drug abusers, drunkards and bullies. There are cases of fist fights, threats to beat up teachers, stabbing each other with knives, expulsion as a result of beer drinking and theft especially by boys.

Two Form 4 boys aged 16 stabbed each other in 2015. One of the boys had been with parents in South Africa and was born to a South African mother. They injured each other and a police report was made. All this is blamed on lack of parental care, control, guidance and discipline despite article 3 (UNICEF 2014) which provides for protection and security of children from all forms of abuse including harmful use or trade in drugs. Children of migrants are unprotected, insecure and vulnerable to harmful use and abuse of drugs and alcohol. They are sought of neglected and marginalised by teachers, caregivers and the community at large.

Lack of financial provisions, anxiety, stress, frustration, too much freedom, extravagance and violence, all are a result of lack of parental guidance. Sometimes migrant parents discourage employed caregivers or relatives from the extended families from disciplining their children. They believe reports from their children about nasty treatment by relatives. Caregivers then neglect the children and allow them to get out of control. Migrant parents sometimes tell caregivers or relatives to keep their officious nose out of their family affairs. Migrants' parents are also to blame for introducing foreign values about the rights of children, for example, allowing children to bring cell phones to school and magazines with pornographic materials. Sometimes children who visit parents during the holidays are in turn also influenced by values that are not compatible with school rules in Zimbabwe.

Although teachers are doing their best to groom migrants' children, the home lack the same force for character building. Such learners are difficult to discipline compared to children of non-migrants. In some instances, teachers end up ignoring them or removing them from school if acts committed are very serious. Teachers' role of discipline, care and safety is defeated. However, not all children display negative behaviour that is incompatible with schooling. Unfortunately, the innocent are also labelled negatively.

The challenges are categorised in order of gravity as delayed fees, shortages of educational materials and foodstuffs, lack of educational assistance, ill-treatment by caregivers and teachers, anxiety and psychological stress and lack of communication and love from parents and caregivers as well as the burden of household work.

Lack of parental care and communication create distant relationship between children and their parents. Parents are strangers to their children. Children and parents do not know each other and are socially detached. Parents' compensatory behaviour is to spoil the children by using money and by taking them illegally for visits in their countries of destination. Parents have a tendency to change cell phone numbers. This results in children and parents not talking regularly thereby creating further remoteness.

An incident occurred in one of the schools, a father who had left for the United Kingdom when the daughter was in grade 6 came back when the daughter was in Form 4 (5 years later) in 2015. They had been seeing each other in photographs. When the parent visited the school and they greeted each other in the school head's office, there was a moment of silence and tears from both for five minutes.

Another incident was when a child was illegally transported to South Africa through *umalayitsha.* The child was involved in a car accident on the way, got injured and hospitalised in South Africa before meeting the parents. He had lost the diary with the phone numbers of his parents. The child knew the school head's phone numbers by head. The school head was by then in Bulawayo on a vacation. The head of the school was phoned and he had to travel to the school about 220 km to get the phone numbers of his parents and the next of kin to try and link the child with the parents and relatives. Teachers view these children as unprotected, experiencing remote parenting and abused in the name of love. Parents use money and materials like cell phones to show love. Sometimes parents cry over the phones to show how much they love and miss their children.

Delayed fees are detrimental to schooling; learners sometimes miss lessons and sometimes fail to meet deadlines for examination fees. School administrators do not accept shortfalls. They are sent home until *omalayitshas* arrive. The methods of transporting financial remittances are also a challenge. If parents send money through the bank to the school account directly, sometimes the exchange rate is very low and it becomes expensive. If they send the money to caregivers, some funds do not get to the school. Learners are sometimes punished by school authorities for the sins of their migrant parents. School authorities seem not to believe that migrants' parents are genuinely failing to pay fees or to send remittances on time. This is evident from the comments they pass to learners. The shortages are sometimes serious for those in low-cost boarding and those who are lodging at business centres and townships. They tend to have shortages at the same time because of the same delivery dates by *omalayitshas.*

Migrant and non-migrant children at school receive same treatment. Teachers do not know the whereabouts of learners' parents hence give them same work in all subjects and have the same expectations of them. School rules are applied to all students the same way. Treating migrants and non-migrants' children the same way is in itself unequal treatment because their living and schooling environments are different. Those with non-migrant parents have their issues quickly responded to or parents come to negotiate with school administrators, while children of migrants phone their parents who may respond the following month. Even in issues of misbehaving and indiscipline, non-migrant children have parents to apologise to school authorities as quickly as possible. Discipline and counselling are immediately implemented while cases of migrants' children are left unresolved for a long time. The school administrators also tend to view migrants as rich people and do not accept payment plans. They are more sympathetic to non-migrant parents whom they view as poor. School authorities also reject foreign currency abiding to the Zimbabwean government currency declarations for use of Real Time Gross Service (RTGS) and do not accept any excuses.

However, migrants' children have developed survival strategies that include strict budgets, fetching firewood and water adequate for a week during weekends to curb on the burden of household chores. In the evenings, they prepare all the meals for the day though this eats on their evening study time. Migrants' children support each other as siblings or friends, lend each other educational materials and conserve the little they have. Older siblings protect and guide younger ones. Teenage boys protect their sisters from vulnerability to males with illicit intentions about their sisters. Boy children take up the protective role of fathers and girls take up care and guiding role of mothers. Migrants' children create friendships with some community members, or members of the extended families they trust though these do not solve all their problems. Some even create friendships with teachers. Not all teachers have negative attitudes towards migrants' children and not all migrants' children view teachers negatively. They learn and copy from such people the skills of survival. Life experiences equip them with funds of knowledge for use. Such skills go a long way although the parents are not physically there to provide the much needed emotional support to their children.

Conclusion

Although migrant parents are concerned with their children's schooling, they are too far away to give practical support in schooling. Learners are alone in this journey. Teachers treat them as if all things are normal, as if the ground is level for both migrant and non-migrant children. The extent of the challenges depend on whether one or both parents are migrants, gender of the migrant or that of the left-behind child, nature of the family of the caregiver, level of education of the migrant and the caregiver. It also depends on the socio-economic status of the migrant and the caregiver, duration of migration, level of communication between migrants and their children as well as relations between the caregiver and the left-behind child. Though learners try to cope with challenges through experiential funds of knowledge sometimes, the challenges associated with absent parents are overwhelming.

References

Beinstern B (2001) The quarterly review of comparative education. Paris: UNESCO. Int Bureau Educ 1(4):687–703

Cortes R (2007) Children and women left behind in labour-sending countries: An appraisal of social risks. UNICEF, New York

Filippa OM (2011) Zimbabwe adolescents experience of their parents' absence due to diaspora. UNISA. Available from uir.unisa.ac.za. Accessed 12 June 2014

Ganju E (2012) A Mothers' Day tribute: The role of women and parenting in girls' education around the world. Available from http://www.huffingtonpost.com/ganju. Accessed 21 June 2013

Gassmann FS, Siegel M, Vanore M, Wandler J (2003) The impact of migration on children left behind in Moldova. Maastricht Working Papers. Maastricht University, Maastricht

Gavriliuc C (2006) The situation of children left behind by migrating parents. GFMD. Available from http://www.google.co.zw. Accessed 27 Dec 2013

Giddens A (2001) Sociology. Cambridge: Polity Press

Kufakurinani U (2014) Transnational parenting and the emergence of diaspora orphans in Zimbabwe. Brill: Africa Diaspora, Leiden

Laureau A, Weininger EB (2003) Cultural capital NAD Educational Research: A critical Assessment. Temple University: Suny Brockport

Luecke M, Stoehr S (2014) The effects of migration in Moldova and Georgia on children and the elderly left behind. Maastricht University: Europe Aid Project, Moldova

McCrossman J (1966) The reading of the culturally disadvantaged. Available from http://www.brockport.edu.sociology. Accessed 19 July 2013

Moll L, Greenberg J (1990) Creating zones of possibilities: combining social contexts for instructions. In: Moll L (ed) Vygotsky and education. Cambridge University Press, Cambridge, pp 319–348

Noveria M (2014) The population of senior citizens in Indonesia: The issues of care-giving in some international migrant sending areas. Jakarta: Research centre for population-Indonesia Institute of Science

Perera V, Rathnayaka MR (2013) Sri Lanka s' missing mothers. A working paper on the effects of mother migration on children. Save the Children, Baticoloa

Save the Children (2014) Left behind, Left out: The impact on children and families of mothers migrating for work abroad. Available from save-sl-leftout. Accessed 12 Dec 2014. https://resourcecentre.savethechildren.net/library/left-behind-left-out-impact-children-and-families-mothers.

Tufis A (2007) The effects of the absence of migrant parents on their children school performances. Soros Foundation. Available from http://www.childmigration.net. Accessed 12 June 2014

UNESCO (2012) Teachers for all: global campaign for education, policy briefing. Collins Bartholomew Ltd, Paris

UNICEF (2008) The effects of parental deprivation on the development of the child left behind by Moldovan migrants. UNDP/CRIC, New York

UNICEF (2014) A summary of UN CRC. Available from Childrenandyouthprogramme.info/pdf.uncrc-summary-version. Accessed 28 Jan 2014

Xiang B (2005) How far are the left behind left Behind? A preliminary study in rural China. Popul Space Place 13(3):177–191

ZIMSTATS (2010) Small area poverty estimation: statistics for poverty eradication. UNICEF and World Bank, Harare

Chapter 6
"Low-Cost" Boarding and "Child Renters": Boys and Girls Schooling in Risky Physical and Social Environments in Bulilima and Mangwe

Historical Context of Physical and Socio-Economic Risky School Environments

The independent Zimbabwe ushered in the education manifesto which was later turned into an Education Act in 1987. The act gave each individual the right to education despite race, religion, ethnicity, disability, gender and creed among others. This coupled with the Education for All (EFA), free compulsory education increased enrolments in few schools that were characterised by bottleneck system of education during the British colonial rule. The situation became worse in secondary schools. Secondary educational infrastructure facilities were scarce. In new resettled areas following land redistribution to employees of white commercial farmers who were formerly denied education, there was dire need of educational facilities.

"Bush boarding" or illegal boarding rose at the same time with satellite or annex classes established in primary schools to cater for students travelling long distances because of lack of schools. Satellite or annex classes were where secondary classes were conducted at existing primary schools. Annex secondary schools were cheaper than building new schools yet ensured easy access to secondary schooling (Mutale 2015). Before the establishment of annex schools, children travelled more than 10 km to school a single journey to the few schools established during the colonial rule. Rural day secondary schools were needed to cater for all secondary students following the education manifesto and education for all at independence.

Bush boarding also rose following the establishment of rural day secondary schools which were also a number of kilometres apart. With the 2000 land reform and redistribution, pressure on educational facilities worsened. There is dire need for educational facilities in new resettled areas and in commercial farms. Previously, employee farmers and their children were excluded from education. In some farms, learning facilities are non-existent or are of poor quality. Schools of better quality are far apart, a two-hour walk a single journey. People are too poor to afford proper conventional boarding schools. Learners from poor families then squat for the term

W. Matsa, *Marginality, Migration and Education*,
https://doi.org/10.1007/978-3-030-60873-6_6

in shacks close to the school or rent a house as learners to avoid travelling long distances every day Dube (2014).

"Bush boarding campuses" are uncontrolled by adults or guardians. Children live in squalid conditions with no sanitary or water facilities. They are not conducive to learning and are risky. A girl was raped and murdered at a bush facility in Manicaland. The Zimbabwe National Family Planning Council in a workshop for adolescence on sexual and reproductive health forum reported that girl children who squat at shop buildings after failing to secure accommodation for schooling are exposed to early sexual activities with older men. About 300 in Manicaland live without parental supervision exposing them to risky behaviours (Chegaza 2018). In Bulilima and Mangwe, most of these children are without parents due to labour migration.

Chigunwe (2018) also expressed rising concern over bush boarding and child renting as sensitive issues, and that access to education is still a challenge to learners in many parts of Zimbabwe. Learners are still forced to walk long distances to school. Bush boarding and child renting are desperate efforts to curb distances of 25–30 km that cannot be commuted every day. However, these expose children to a lot of temptations and risks.

Many children depending on the region and place are raised to endure the agonies of seeking the future linked to education (Muleya 2019). The probability of children not continuing in school is twice as higher in rural schools categorised as marginalised. In these schools, girls are twice likely to fail to progress in school than boys. In these marginalised rural areas, learners face different levels of inequality that depend on the dynamics of gender, power, age, and social conditions at home (Care, 2014: IGATE broad gender analysis report. Unpublished).

Most rural children in remote areas and resettlements have no access to schools. They travel long distances. The normal walking distance to school is 3–4 km for those aged 8–16 years old. Matabeleland South province among other regions is most affected in terms of proximity to school, accommodation for schooling and quality education in general. There are more dropouts during transition from primary to secondary education. Incidences of child early marriages, juvenile delinquencies are common in places where bush boarding school is practised.

In rural districts in natural regions, 4 and 5 where Matabeleland South is situated shortage of school and long distances travelled affect especially girls' attendance and exposed them to abuse on their way to and from school. This ecological region has low to moderate rainfall causing food shortage as barrier to education. Attempt is made to create equitable learning environment in marginalised groups to access education. To curb long distances, bicycles were distributed to a ratio of 30 boys to 70 girls shifting traditional gender norms to improve girls' learning, attendance and retention (Chigodoro 2017). The unfortunate part, however, is that migrants' children in most cases are not part of the selection criteria of the marginalised groups even if they are in a worse situation than non-migrants' children. In this part of the country, a bicycle is a common mode of transport bought and sold by migrants. The assumption is that migrant parents should buy bicycles for their children but not all migrants remit sufficiently. Some do not even remit.

In Beitbridge Matabeleland South Ward 14, children trek distances of between 10 and 17 km to school. To ease long distances, a total of 44 pupils stay at Jopembe primary school. Under commonly practised bush boarding. During the day, the makeshift classrooms are used and at night, the same classes are dormitories. When one gets into the school, he or she is welcomed by blankets and clothes on the fence. One is greeted by exhausted faces of learners from not so comfortable sleep. Children are crowded and sleep on the floor. About 24 boys stay in one classroom with male teachers while 20 girls stay in another with female teachers. Kageler (2015) reveals that 2056 new schools need to be constructed countrywide. Additional 33,636 classrooms are needed and 3554 classrooms are derelict and dangerous. Alternative building materials and methods need to be thought out to meet the schools functionality standards. Besides shortage of accommodation and facilities, girls are vulnerable to sexual abuse. Six pupils have reported cases of rape from the above school in the past 3 years. The long distance places feeding the school are Mazunga (11 km), Tshambengedzani (12 km), Mpanyani (18 km), Shonisani (14 km), Section 3 (9 km) and Ghova (17 km) (Muleya 2019). The enrolment is low and the school levies cannot build classrooms and teacher's cottages. There is one block of classrooms built by the Zimbabwean military army in 2016. The situation is dire; the community is mostly made of low income and the retired.

A woman is hired at Jopembe bush boarding school to prepare meals but the living conditions are disheartening. Parents try to put resources together for teacher's cottages and ablution block facilities. The toilets are on loose soils and are giving in. They can collapse anytime. The member of parliament for Beitbridge confirmed that the situation was deplorable and needed urgent attention. The area has a population estimate of 200,000 to 15 secondary schools, out of which only two are boarding schools (Muleya 2019). The educational facilities are not conducive for learning. In such situations, migrants' children suffer twice as much in issues related to accommodation and security.

The state of schools, distances and shortage of accommodation influence teachers to shun these schools. Kageler (2015) points out that generally, there is serious shortage of teachers in Mabeland North and South with the North having the highest number of unqualified teachers. There is inequitable distribution of teachers, 86% in wealthiest and 61% in poorest districts. Teachers shun poor provinces and districts. In poor districts, such as Bulilima and Mangwe, 10% of teachers have less than 10 years of experience.

In Bulilima and Mangwe, bush boarding still exists although teachers and administrators do not openly acknowledge. Some of the formally bush boarding accommodation has been transformed to low-cost boarding schools. However, there are remnants of children who still squat in old buildings and homes near the schools or within the school premises to curb distances. These learners live without direct adult supervision. Learners and members of the communities are aware of these. Parents or guardians with money send their children to low-cost boarding which are cheaper than conventional boarding schools.

Low-Cost Boarding Schools

Low-cost boarding schools are rural day secondary schools providing less formal boarding facilities to learners at a low cost. There are about three low-cost boarding schools in Mangwe and three in Bulilima. Some are continually being upgraded and expanded to make them formal or near-formal. The low-cost boarding are legal, known by the ministry of education and are supervised by the ministry, despite that it is an endeavour by parents' associations to provide affordable accommodation to curb distances and costs charged by mission and government boarding schools. There is a government circular that regulates how they should be run by school development association of parents. The government does not fund low-cost boarding schools, and there are no non-governmental organisations that sponsor them. However, the government does not take total responsibility but only play a regulatory role. Parents also cannot afford the standard provisions as these are funded by poor parents who cannot afford conventional boarding schools. Lack of funding becomes a serious problem. Parents charge affordable levies they see fit for approval by government.

Accommodation, infrastructure and equipment for use is sourced by parents. Four of these have reliable electricity supply for lighting system, excellent water and other amenities making them conducive for learning. The ministry of health sometimes carry out inspections conducted by Environmental health Technicians (EHT) or Environmental Safety Technicians (EST). Low-cost boardings do not have health centres or clinics as is in mission and government boarding schools. Schools also have no formal security by way of security guards, alarm systems or electric charged wires. Theft is very common.

The majority of learners in low-cost boarding in Bulilima and Mangwe are children of migrants who cannot afford government or mission boarding schools. The children are taken care of by school superintendents or matrons. These are local people with no qualifications of being board masters and matrons. However, they are formally selected through interviews and a criterion of their behaviour, care, minimal education and social standing in the community. The cooks are also locals from the community who also do not have qualifications as cooks. There are no resident pastors or chaplains as is the case in mission schools. Guidance and counselling is done by teachers.

The facilities are affordable compared to conventional boarding schools. The cost is made cheaper by the fact that sometimes groceries and toiletries which are cheaper in South Africa and Botswana are sent as goods than money. Learners pull food resources together. They bring varieties of food from home fortnightly or monthly.

The rationale for the emergence of low-cost boarding is that in the two districts; there are no government boarding schools but only mission boarding schools which are expensive. Mission schools accept students on payment plans and give first preference to learners from their denominations. In Bulilima, there are Methodist, Seventh Day Adventist and UCCSA (traditionally known as London Missionary

Society) mission boarding schools and in Mangwe there are Roman Catholic, Anglican and Salvation Army mission schools, some of which are single sex schools. Low-cost government boarding schools disregard gender, faith and class giving all learners equal opportunities to access education.

The schools with low-cost boarding have the majority of learners commuting from home every day. Other learners are from nearby townships and business centres where they stay with caregivers who own or manage businesses in the townships. Some stay with government workers in police stations, hospitals, clinics, Agritex centres, rural district development offices, post services and veterinary services among others.

Failure to get a suitable female caregiver and distances to school force migrant parents to opt to take the child to a low-cost boarding or to lodge in houses at the business centres. Most of those in mission boarding schools are those who have either parents outside or parents with better jobs in destination countries. Some parents initially travel with children to the diaspora. Some children are then sent back by parents to do their education in Zimbabwe. Most of these pose challenges of care when it comes to walking long distances to school; diet and chores hence find it difficult to stay with relatives or caregivers.

More girls than boys are accommodated in low-cost boarding for security reasons and distances to be travelled to and from schools. The belief is that boys can travel long distances from home. However, the challenges are that only a few can be accommodated in these low-cost boarding schools. Some schools do not have such services, those that do, can accommodate only a few students because of inadequate facilities. Community people expect the government to come in to assist schools to expand such services.

Being less formal has its own disadvantages, in 2015; some boys were expelled in one of the schools in Mangwe low-cost boarding for beer drinking and for threatening to beat up teachers. Low-cost boarding have no resident pastors or chaplains for guidance and counselling, instead, pastors from mission schools are sometimes invited to visit and counsel learners. Lessons in most cases are not specific to the challenges of migrants' children but general topics for all the learners. The counsellors do not focus on problems of children with absent parents.

Learners in low-cost boarding lack communication and attachment with parents. During the holidays, migrant parents come for a short time during the busy season for cultivation, which is December. If children visit their parents in the diaspora, parents spend time at work. Boarders are not allowed to phone their parents directly but phone through school heads, matrons and superintendents. This is done when there are important issues to be discussed, for example, fees and shortage of educational materials. Learners are not happy with this lack of communication. They do not trust people and are denied chance to confide to their parents. Although these living conditions are similar to those in mission boarding, non-migrants' children in mission boarding have parents or guardians who visit during prize giving days, consultations and annual general meetings.

The other challenges experienced by learners with absent parents are health-related issues. If those in low-cost boarding are ill, they are treated at local clinics which, in most cases, have no medication. Migrant parents are far away and cannot be expected to buy the medication needed. As a result, children suffer a lot. For serious illnesses, they are taken to referral hospitals like Plumtree and Brunarpeg general hospitals. Most migrants' children have no medical aid schemes and the next of kin caregivers in Zimbabwe have no cash to pay for travelling expenses, admissions and medications. In most cases, relatives also cannot pay medical bills or attend to them regularly in hospitals until migrant parents send money. Some are sent home to relatives.

"Child Renters": Children Renting Accommodation for Schooling

Much of research on children has been on vulnerability and risks ignoring the fact that children have means of acquiring new knowledge for survival (funds of knowledge). Children are competent, resourceful and knowledgeable. Children can build and perpetuate their lives and education while recognising their conditions of extreme challenges (Kendrick and Kakuru 2012). In Uganda's Rukai district, children from child-headed households learnt from competent members of society and peers to survive on their own. They supported each other and families continued even in the absence of parents. Orphanage families evolved to meet the needs of its members. Orphans developed adaptive capabilities through experiential learning from life challenges.

Vygotsky (1978) concurs that children can acquire knowledge that helps them to adapt through social interaction with significant people in their lives (shared knowledge). Children learn patterns of knowing and doing in formal and informal settings. Migrants' children in low-cost boarding and those renting for schooling in Bulilima and Mangwe also learnt survival skills from peers and community adult members. In Uganda, children in child-headed households were able to make meaningful decisions, looked after young siblings and had network of support. They negotiated for resources, did self-advocacy and helped young siblings with school work, budgetting and the distribution of the little resources they had, even resources for schooling. Orphaned children living in child-headed households sold small items for cash or did barter exchange. Children acquired knowledge across a wide range of community members and demonstrated a strong ability to cope with life's challenges. They had practical management skills, a sense of independence and even strength to cope with stress. They also had the ability to make important decisions, could work hard social networking and mobilising social groups for income generating activities. Children pulled their resources and experiences together. Children mentored themselves and constructed positive social realities of their lives (Kendrick and Kakuru 2012).

Migration in Bulilima and Mangwe gave rise to vulnerable child renters who face challenges of absent parents due to migration. Children are sent financial remittances to rent houses in townships, growth points and business centres for proximity to secondary schools. These children enjoy total independence in all spheres of life. Some of the learners rent houses in the township owned by migrant property owners and in most cases, the payments for rents are in Rands. Unlike in low-cost boarding, renting learners live alone with no security or supervision by an adult. Their living conditions are poor. Some experience sexual abuse and are insecure and vulnerable to drug and alcohol abuse. They drink and influence others to drink. They mix coca cola with pills to turn it into beer. They witness fist fights by drunkards who sometimes stab each other with knives. Learners learn violence and bad habits.

Such experiences make children vulnerable to violence and other forms of abuses like prostitution. In these growth points, townships and business centres, it is noisy and not conducive for learners, socially and academically. There is prostitution in night clubs where radios are on full blasts till mid-night making the places not suitable for studying.

The girls who live in rented houses who have remittances delivered by *omalayitshas* monthly are viewed as girlfriends of these transport operators. They are also suspected of going out with police officers from nearby camps. Girls are infuriated by these suspicions. Both boys and girls are mocked by teachers as *omalayitshas'* children and boys as future *omalayitshas* in waiting.

Those lodging in townships also have health challenges. If they suffer from minor illnesses, they are cared for by friends. Friends prepare food, water for bathing and leave them alone in the room for the whole school day. This happens every day until the learner recovers. They nursed each other. Sometimes they escorted each other to clinics or to relatives. This was observed by Ankut (2014) that the right to education and other rights of children are violated every day. In this part of the country, indeed, childrens' rights are violated every day.

Children lack health care and medical advice (UNICEF 2008). Most health problems are related to poor habits of eating, hygiene, abuse of medication and drugs. This was confirmed by UNICEF (2008) in Moldova that the quantity and the quality of food is poor either because of lack of it or as a result of poor preparation. To those with money, there was tendency to buy fast food products which are also poor nutritionally. The children rarely had three meals a day. Lack of medical aid schemes coupled with lack of advice on personal health, stress-related problems and close persons to share with intimate issues, tended to hide health problems. Most of their problems related to personal health, eating and bathing habits, pregnancy, sexually transmitted diseases, HIV and AIDS and general reproductive health. The situation of poor diet was also found to be worse in the Chinese provinces of Hubei, Anhuix, Henan and Si Chuan for those children living in shell villages (slums) alone. Children ate poor quality food, rotten and poorly cooked. Children prepared meals for themselves.

In Bulilima and Mangwe, children are nicknamed *phekeyakhe* meaning those who cook a small pot for themselves or *chunks* because they sometimes rely on soya meat popularly known as chunks as the only relish. Preparing meals for themselves has implications on their labour, and time for school and study. Quality food especially feeding times and lack of balanced diet directly affect schooling. The majority of migrants' children in Bulilima and Mangwe do not have refrigerators in their rented houses. Some depend on shopkeepers. Not all have access to refrigerators and it depends on relations with shopkeepers.

Those living in incomplete houses have no access to electricity. In most cases, their food gets bad especially during the hot season. Because of strict food budgets, if the food is no longer fresh, they cannot throw it away. Lack of time forces them to prepare all meals for the following day in the evening before. Every lunch they eat cold food and it is unhealthy. The lunch is usually cold thick porridge made of maize meal, millet or sorghum with chunks, beans, matemba, *(mfushwa)* dried vegetables and (*macimbi*) mopani worms as relish. For breakfast, they have tea without milk, accompanied by charcoal flour backed bread and (*maqebelengwana*) boiled balls made of mealie-meal. Most of the food is starch with relish which is not nutritious, for example, dried vegetables and poor quality chunks. Those with parents with better remittances may also have tinned fish, tinned beans and milk. These sometimes have rice, spaghetti and macaroni. Lunch time is only an hour including time for travel to the house depending on the distance. They cannot make fire. Some carry lunch boxes with cold food. Generally, because of peer pressure and fear of being mocked, adolescents are not in the habit of carrying lunch boxes if the food is of poor quality. Some end up preferring to forfeit their lunch or carry junk food which is not filling. Examples are things, Jiggies; (*maputi*) popped maize meal, biscuits, sweets, diluted drinks and Jupiter willards, that is, if they have money. Lack of money also drives girls into prostitution to get supplements. Because they stay alone, no one controls their behaviours.

The houses they rent do not have security fences and there are drinking places of entertainment. Girls are also influenced to drink by boys and, in addition, experience sexual abuse and violence. Exposure to sexual activities at an early age affects their academic performance, retention and advancement in schooling. Some girls are impregnated or beaten up by *njivas* and other non-migrant people of the community. They learn bad behaviour from the community and from those who once went to South Africa. Those who are lodging are also looked down upon as poor because their parents cannot afford even low-cost boarding.

Sometimes their houses experience burglary. Their properties, groceries and gadgets are then stolen. The living conditions are poor. They share toilets with vendors, business caretakers and drinkers. Those renting incomplete houses in townships have their own challenges. Some of the incomplete houses do not have electricity and tap water. They sometimes drink untreated water from dams, wells and rivers as community people do. The life of child renters cannot be compared to that of children who stay with their mothers, grandmothers or those living in mission, government boarding and low-cost boarding schools.

Inadequate water and sanitation is a key factor affecting girls' learning. They often remain at home during menstruation if water situation is dire at home and school (Chigodoro 2017). World Vision (2014: IGATE baseline report. Unpublished) also observed that 19% of the out of school girls feel that home and schooling environments have great impact on their schooling. About 25% colluded that issues of menstruation and sanitary hygiene management made girls to develop a negative attitude towards school. The schooling environments must be gender responsive, conducive for both male and females. If schooling environments run short of water, it is very difficult for girls to attend school regularly especially if there is also water shortage in the school vicinity. At school, child renters and those commuting from home do not access facilities for conventional boarders and low-cost boarders in halls of residence. Time taken to fetch firewood from the forests and water from dams, rivers and wells in Bulilima and Mangwe taxes on their time for schooling. Girls' security is also at risk as they expose themselves to men who prey on them.

Despite various challenges encountered by migrants' children, they demonstrated capacity for resilience. Renting to reduce distances travelled to access schooling was itself an adaptive strategy by both migrant parents and their children; otherwise, children would not access secondary education. Rented houses resembled family accommodation and peers became siblings. Children created among themselves networks for support similar to family relationships. In case of illnesses, they were cared for by friends. Friends prepared food and warm water for bathing. They nursed and escorted each other to the clinics.

Although child renters lived alone without adult guidance and supervision, they made efforts to have at least three meals a day though in most cases the food is substandard and low nutritional value. They did financial and food budgeting of alternative meals and various types of relish they got hold of. They were assisted by shopkeepers to conserve and refrigerate perishable food items. Those with no access to refrigerators relied heavily on dried foods, for example, chunks, beans, mopani worms, matemba as relish. Other non-perishable food items were biscuits, maputi (popped maize meal), jupiter willards, drinks, cereals among others. Some ate heavy thick porridge and relish in the morning before going to school and forfeited lunch if they had nothing to carry as lunch.

Migrants' children braved the stigma of being poor, living alone as orphans and some in dilapidated and incomplete houses just to access education. They worked hard combining work with schooling. They fetch firewood and water from rivers, wells and boreholes. In times of shortages, girls improvised on sanitary wares and conserved the scarce water for use during their menstrual periods. Their life was poor in comparison to those in low-cost boarding and those commuting from home but they made tremendous effort to adapt to the challenges through use of skills from the funds of knowledge, knowledge acquired through experience of living in difficult circumstances.

In comparison to children of non-migrants, migrants' children are more independent, hardworking, relatively mature, budget conscious and strong willed. They are responsible when it comes to home management and decision-making. However, challenges of poverty, absence of parents and reliable caregivers as well as convergence of economic, educational and social marginality compromise their potential of resilience.

Conclusion

Although the rise of low-cost boarding and child renting rose as an attempt to curb problems associated with lack of access to schooling due to distances and cost of government and mission boarding schools, their informal nature has worsened the problems of learners with absent parents in Zimbabwe. There are worse challenges associated with distance from caregivers, migrant parents, renting and living in a low cost without a responsible guardian to take care of problems related to care and schooling. The educational infrastructure leaves a lot to be desired and lack of resources limit the capacity of migrants' children to cope with challenges. Child renters live in difficult and risky circumstances not conducive for quality learning.

References

Ankut PY (2014) The African charter on the rights and welfare of the child: linking principles with practice. Available from http://www.hurisa.org.za. Accessed 15 Jan 2014

Chegaza S (2018) Bush boarding houses expose girl learners. The Manica Post 28 December 2018. Manicapost.co.zw/bush-boarding-houses-expose-girl-learners/ Accessed 10 October 2019

Chigodoro U (2017) FAWE conference on girls' education in Africa 23–25 August 2017. Care Zimbabwe, Lusaka

Chigunwe D (2018) Government frets over bush boarding facilities. The Herald 6 April 2018. https://www.herald.co.zw>gvnt-frets-over-bush-boarding-facilities/ Accessed 26 March 2020

Dube S (2014) Poor conditions stumbling block to children s' learning. The chronicle 3 March 2014. https://www.chronicle.co.zw/poor-conditions-stumbling-block-to-childrens-learning/ Accessed 25 March 2020

Kageler S (2015) Education sector analysis Zimbabwe: final report. https://www.globalpartnership

Kendrick M, Kakuru D (2012) Funds of knowledge in child headed households: Ugandan case study. Childhood 19(3):397–413

Muleya T (2019) The agonies of seeking fortune for a rural child. Available from https://www.chronicle.co.zw

Mutale Q (2015) Challenge facing children in rural Zimbabwe: a case of Tyunga and Luunga words of Binga district. Res Humanities Soc Sci 5(9):32–37

UNICEF (2008) The effects of parental deprivation on the development of the child left behind by Moldovan migrants. UNDP/CRIC, New York

Vygotsky LS (1978) Mind in society: the development of higher social processes. Harvard University Press, Cambridge

Chapter 7
Children Belonging to "No One": Learners Trapped in Strained and Suspicious Relationships

Parent–Teacher and Community Partnership in Schooling of Children

Teacher–parent and community collaboration is very key for quality education for children. The teacher–parent partnership has the common goal of educating the child; however, today the rights and obligations of the teacher are not clearly defined in practice (Osthuisen 1992). The role of teachers has become specialised, purposeful and professional although their other role is to educate learners on norms and values of society. The teacher is expected to educate, discipline and provide custody, duty care and safety environments during school activities in the classroom, laboratories, playgrounds, sports fields, swimming pools and trips among others. The physical, academic, social and economic problems that emanate outside the school are outside the teachers' jurisdiction. This leaves migrants' children in a dilemma particularly because at home they do not stay with their biological parents but caregivers who neither care about such problems. The role of the school and that of the family has been categorised and dichotomised limiting family to nurturing at home and school to classroom instruction instead of partnership in holistic concerns of the child. This disrupts possibility of effective partnership and collaboration in schooling of children especially migrants' children.

Under *loco-parentis* principle, teachers have a legal and moral responsibility to care after students, and this power is transferred from parents to the teachers (Worley 2003). In education, this becomes the basis for authority. In addition to his teaching, the teacher has the power to discipline, control and care to ensure safety and well-being of students taking into account home, school and community factors affecting children. Worley (2003) summarises the functions of a teacher as maintenance, protection and education. To Ngwana cited in Nakpodia (2012), the role of the teacher is that of instruction, supervision and safety.

One of the errors and omissions of the twentieth century is to hold the traditional view of the 1960s of treating parents as "silent" or "non-existent" partners, where

W. Matsa, *Marginality, Migration and Education*,
https://doi.org/10.1007/978-3-030-60873-6_7

the child is the only connection between the home and the school, parents and teachers (Yawkey and Bakawa 1974, p. 2). Teachers were to do the caring, teaching and guiding. Caring for teachers was assumed to be totally independent from the child's learning at home and learning had nothing to do with the child's rearing practices. The home and the school were two separate environments. This gap is evident in Bulilima and Mangwe in that teachers do not realise that parents share the responsibility with teachers in care and education of children and that their influence cannot override that of parents. Parents, children and teachers should act and interact as a unit for quality care and educational outcomes. Caregivers also do not realise that their role as stand in for parents. They think their care is totally independent from that of teachers and that the school and the home are separate. Caregivers do not see the importance of joint participation of home and the school in a harmonious manner.

There is growing need in Bulilima and Mangwe for the emergence of the triad of participants in care of migrants' children or non-migrants' children. Schooling, childcare programmes and services as well as staff must be committed to the concept that the child's parents are the most important persons in the child's life at home and school. Parents and caregivers are valuable assets and must be genuinely involved in children's programmes (Yawkey and Bakawa 1974). Parents share the responsibility for care and education. Home, neighbourhood and the classroom make up the learning environment. Teachers are to develop active interaction between home and the school. Parents and staff help to give security, love, information, control and direction to build character behaviours congruent with quality schooling. Teachers are parents at school and parents are teachers at home. Ignoring the triad participants of care result in neglect of children.

With the above responsibilities for teachers and parents without dichotomising the family and school environment, children left behind by migrating parents would be very safe. However, Nakpodia (2012) points out that this is not what practically happens on the ground. It was discovered that in Nigeria, the teachers' role of discipline, care and safety was lacking in both rural and urban schools. This role is also lacking in Bulilima and Mangwe. This is particularly pathetic for migrants' children because at home also no one does that role because of absence of parents due to international labour migration. As a result, these children belong to no one and are not attached to anyone.

In the Caribbean, Baker et al. (2009) discovered that children had a feeling of abandonment, neglect, low self-esteem, anger and material obsession. Children were attached to material resources. They were full of despair, lack of trust and the inability to concentrate. Children spent their entire lives battling with feelings of rejection and loss. Further broken promises of reunion with parents tend to result in emotional instability and distrust with anyone.

Cortes (2007) found that in the Philippines lack of communication worsened the psychological well-being of migrants' children. About 90% of the parents had left children nine years ago. Half the mothers had not returned for 4–6 years and 10% of the children had not communicated with parents. Only 16% of the children had daily communication with parents. In China, cell phone communication was low, infrequent, twice a week and taking shorter times as three minutes because of expenses of the cell phone (Wei Lu 2011). The parents enquired more about performance and not well-being. Teachers were also concerned with performance at the expense of well-being. Caregivers cared less about both performance and well-being. Parents gave school performance first preference and sometimes would not even talk to their children but caregivers. Parents depended on reports by caregivers who sometimes misrepresented facts. This created friction not only between children and caregivers but also between parents and caregivers. Less than 20% of the children shared their problems with caregivers and 30% kept problems to themselves. Quality of communication interchanges with parents was very low and limited to instructions concerning use of remittances in Ecuador (Cortes 2007). Children's well-being was neglected by both parents and caregivers. This is the situation of migrants' children in Bulilima and Mangwe.

Learners cannot turn to teachers for help on issues to do with well-being. FAWE (2006) discovered that in Africa, generally learners found it difficult to seek guidance from teachers on personal issues because they view teachers as authoritarian, unquestionable, unapproachable and distant. Teachers do not see it as their responsibility to attend to learners' personal issues. They do not bother to understand the behaviour and plight of learners. Teachers associate boys with arrogance, lateness, drunkenness and address them harshly. They ignore issues of sexual maturation and adolescent problems (period of storm and stress) for both boys and girls. They down play the fact that at this stage learners are full of stress, confusion and indecision.

In Harare, Zimbabwe, Filippa (2011), adolescents had feelings of abandonment and rejection. They had conflicting feelings of gratitude that their parents provided for them well but at the same time angry with parents for abandoning them to caregivers. Family restructuring and adjustments were stressful life events to adolescents. Children had no one to openly cry to, had tearful moods, felt sad and empty. Friends and relatives did not compensate enough for parental absence. They lacked parental advice and showed irritability when they spoke to caregivers. To them home was the "the four walls in a durawall" with no warmth of a home.

In Zimbabwe, high-density suburbs of Masvingo, migrants' children did not have good relations with guardians and housekeepers meant to be their caregivers. The caregivers sometimes took instructions from migrant parents and imposed on children as they are, for example, keeping them inside durawalls, denying them visits and being too strict for comfort. They wanted to go shopping and needed one or two friends or school peers. Other caregivers did not even report children's misbehaviour because they wanted to remain in favour of children and their parents for the continuity of their jobs at the expense of the children's well-being. They feared that if they complain or report to remitters, their salaries or remittances would be cut. Even relatives preferred to keep migrants' children for material benefits. In such instances, distance parenting does not provide adequate care arrangements because children do not benefit in terms of well-being.

Educators cannot do it alone. Family, school and community partnerships are one of the biggest problems facing our nations. Schools work in isolation yet the whole community has an essential role to play in the growth and development of its young people (Von Roekel 2008). This vital role results in high-quality education for all students. Some parents and community members still view student learning as the responsibility of educators.

Family–school collaboration is key in activities involving school staff, parents and the community (Department of Education Employment and Workplace Relations 2013). Mothers, fathers, relatives, grandparents, foster parents, business people and caregivers as well as staff make up members of the collaborative partnership. The staff, family and the community should work together. This is lacking in Bulilima and Mangwe in relation to migrants' children. There are no cordial working relations between school (teachers), families (parents, foster parents and caregivers) and the community at large. As a result, parents or caregivers know very little and care less about the needs of students at school. The teachers also do not know about students' needs at home and school. Parents and caregivers need to share ideas with teachers about learners.

Families and the community are first and continuous educators influencing children's learning and development during school years and long afterwards (Department of Education Employment and Workplace Relations 2013). Schools, therefore, are to recognise the primary role of the family in education, hence collaboration and partnership. Effective schools have high levels of parental involvement. Family involvement is central to high-quality education and is part of core business of the school. Each part should be viewed as making a valuable contribution. Different contributions are to be respected and valued. Unfortunately, in Bulilima and Mangwe children or learners belong to none in the trialogue of teachers, parents, and caregivers. The trio does not view learners as a shared responsibility. Teacher–parent partnership is lacking in the education and care of migrants' children. Parents and caregivers should be educators and nurturers at home while teachers educate and nurture at school.

According to Von Roekel (2008), parent (family), school and community partnership correlates with academic performance and school improvement where the three work together to support learning. Students earn higher grades, attend school regularly, stay in school longer and enrol in higher levels. Educational aspirations are fostered. The teaching, learning, and provisions of students be it in resources or social services, it all depends on the family–community–school partnership. The three need to participate and collaborate in all school-oriented activities.

Lack of parent–caregiver and teacher collaborative partnership in Bulilima and Mangwe community is evidenced by suspicious and strained relations, blame shifting, neglect of children (learners) by the school (teachers), family (parents and caregivers) and the community at large.

According to the Department of Education Employment and Workplace Relations (2013) family–school partnership is based on trust and mutual responsibility. Unfortunately, these are lacking between teachers, parents, caregivers and learners. Suspicious and strained relations show lack of trust. Building a culture of trust is an important element for family–school partnership. Karakus and Savas (2012) discovered that high trust levels result in increased partnership and mutual responsibility. Low levels of trust of parents by teachers create low-level trust of students by teachers. High trust levels also lead to high interest, concern and cooperation with the other party. Low trust climate result in suspicion, pretense and lack of genuineness, often leading to conflicts.

Trust improves parent involvement and makes teachers develop humanistic rather than custodian and dominating strategy in control of learners. They deal with learners in humanistic perspective and not in a forceful manner as if dealing with objects. They also work with parents in learners' discipline viewing also cases from learners' and parents' perspective. There is a reciprocal relationship when parents get involved more in learners' education. If parents are less involved, teachers do not develop trust for parents and students. Trust between people and their institutions or organisations improve solidarity. In Bulilima and Mangwe, teachers do not trust migrants' parents, caregivers and migrants' children. Migrants' children do not trust teachers and caregivers. Caregivers view teachers, migrant parents and teachers with suspicion. The relations are strained to the extent that there is no cooperation or mutual responsibility between family, caregivers and teachers. Relationships are not genuine. This results in neglect of migrants' children. There is no parental involvement in schooling of migrants' children and teachers ignore nurturing migrants' children at school.

Dodging of Responsibility by Migrant Parents, Caregivers and Teachers

Responsibility and blame shifting is evident when the trio fail to honour important school calendar days like annual general meetings, consultation days, prize-giving and visiting days. Migrant parents shift the responsibility to caregivers who think that, it has nothing to do with them. If asked to attend, caregivers do not treat the issue with the seriousness it deserves. This creates a distance between migrant parents and the school authorities. Migrant parents depend on unreliable information from the few caregivers who sometimes attend. There is serious lack of educational assistance from both migrant parents and caregivers. The schools do not take into consideration the situation of learners with absent parents. Children belong to no one as far as schooling and educational support is concerned.

Migrants' children also experience the challenges of strained relations between them, teachers and caregivers. Teachers ignore the role of guidance and discipline as loco-parentis (Nakpodia 2012), limiting their role to tutoring and instruction on academic knowledge in the classroom. Teachers feel that guidance and counselling on issues relating to absent parents are not their responsibility.

Blame shifting is evidenced by the trio's perspectives of challenges of migrant children. Caregivers confirmed challenges they experience themselves in the upkeep of migrant children citing burdens shouldered on them by migrant parents and their children. They feel over-stretched by the expectations of migrant parents and their children when they have their own children to look after. To them, migrant parents leave them with the total responsibility of their children, making it appear as if caregivers are discriminatory, unloving and uncaring.

One of the aunts complained that the foster child she cared for thinks that she is discriminating because she paid and bought everything for her own children in time. The foster child expects her to do as she does for her children including taking the foster child to South Africa. She sometimes go to South Africa with her children during the holidays because her husband works there. The foster child has no passport. The foster aunt used to do some of the things for the foster child before the migrant mother got employed. She withdrew some of the privileges when the migrant mother started working. After getting employed, the migrant parent continued to push responsibility to the foster aunt. In aunt's words, (*bayasisunduzela abantwababo*), that is denying and pushing responsibility to foster caregivers.

When migrants buy goods, they buy only enough for their children. As a result, their children end up getting double shares if caregivers try to cover up for the delays and shortages. Some migrant parents are selfish, buying only those things that benefit their children directly, ignoring basic needs. They send clothes and educational materials like uniforms, fees and stationary forgetting that their children need food also. Caregivers sometimes end up selling clothes of migrants' children in order to get mealie-meal and relish.

A 78-year-old grandmother complained that her migrant children send fees ignoring, stationery and emergencies. Children end up copying notes from others very late or using one exercise book for exercises and tests. Migrants also ignore pocket money for sports, trips and civics days.

Another caregiver, a foster parent who is a teacher, complained that migrant parents burden even people who are not blood relatives by expecting them to meet their children's school needs and food. The intervals for remitting are spaced and migrant parents focus on needs such as drinks, cereals, surf, soap, toothpaste with no rice, flour, mealie-meal, cooking oil and relish which are basics for the child's survival. Relatives of the foster child also keep quiet leaving everything to the foster parent to the extent that foster parents end up using their resources. Foster parents are forced to take total responsibility, the child reports all the shortages to the foster parent even emergencies, forcing foster parents to go out their way to curb the child's sorrow. This is viewed as burdensome by caregivers.

Some of the migrants rarely send anything. Most of these are *imitshoni* (those who have abandoned families. They keep quiet throughout the year, leaving caregivers to suffer. When phoned, they do not respond or do so after a long time after several reminders. Incidents of children delaying entry into form one or lower sixth where full uniforms, fees and levies are expected are very common. Some children end up doing piece jobs in shops during the weekends and holidays to access schooling.

When shortages become serious at home, the relations with caregivers become sour to the extent that caregivers tell migrant children that "(*awusomntanami)*", you are not my child. Children have nowhere to report to but have to wait until *omalayitshas* arrive.

A grandmother narrated how she is suffering with her two grandchildren whose parents are migrants. She herself has two children who are also going to school. She struggles to pay fees. Her migrants' children who are parents have never paid fees for their children. Another grandmother, a pensioner, complained that her migrant son was useless. The daughter who used to help her with fees is now married. She is sometimes forced to borrow money and other items which are returned sometimes after threats from lenders. Migrants plunge caregivers into debts which keep accumulating and straining relations with even neighbours and other relatives.

Boys lack close father figures and are spending time in shops and neighbours' houses socialising with men of the community. Those, whose parents never return, think them dead. Culturally in these communities when a close relative dies, very young children are told a lie that the relative is in South Africa, and one day the relative will come back. Those who never saw their parents think they are dead. Boys are particularly difficult to control by females. They want too much freedom and to know where their fathers/mothers are, further straining relations between them and caregivers or parents. The lack of presence of a father as a disciplinarian and a role model creates too much freedom and bossy attitudes. Boys propose a lot of girls, play soccer and ignore school work. They need a strong fatherly hand. There is a vacuum when it comes to boy–father relationship in this community. That responsibility has no one to shoulder.

Mothers also complain that fathers burden them with everything from parenting, budgeting, provision of food and educational support. Sometimes children show signs of emotional stress and think the rare contact with fathers indicate an impending divorce. Girls with absent mothers lack motherly love which manifest in closeness with matrons and treatment of matrons as mothers. They play mothers' babies to matrons expecting matrons to treat them tenderly, as mothers. Caregivers do not cover the gap of absent mothers and fathers because of uncordial relations.

There are strained relations between children and caregivers, caregivers and migrant parents, teachers and migrant parents. This further marginalises migrants' children. Poor foster relationships are not conducive for the education of these children. Of interest to note is serious finger-pointing. Teachers blame caregivers who blame migrant parents, teachers and the learners. Migrants' children think that school authorities, teachers and caregivers are all at fault. The children feel unwanted and burdensome to teachers and caregivers. Migrants' children are neglected and caught up within a trialogue of parental, teacher and caregivers' power of neglect and marginalisation.

Absence of parents seems to produce two different kinds of children: too mature for their ages, or too reserved, isolated and detached. The other group are those who take time to mature, have excessive pride, ill-socialised, bullies with tantrums and signs of stress. This manifests in being violent, abusing alcohol and drugs, stubbornness, and hostility, especially boys. Such behaviours further strain relations with caregivers and teachers and consequently with migrant parents.

Suspicious and Strained Relations

One of the most salient educational challenges of migrants' children is suspicious relations of the quadruple, teachers, learners, caregivers and migrant parents. These strained relations are not always openly observable but in most cases manifests in hidden experiences of migrants' children. Good community, school and home relations create a favourable and quality environment for schooling as pointed out by Thamarasseri (2017). The diagram below illustrates strained and suspicious relations (Fig. 7.1).

In R_1, teachers strongly think of migrant children as rude, disrespectful, directionless, bad, careless, drunkards and parentless while children see teachers as cruel, hateful and jealous. In R_2, caregivers also feel migrants' children are disrespectful, burdensome, uncontrollable and unthankful while children accuse caregivers of being cruel, remittance opportunists, unloving, uncaring and unhelpful. This manifests in hate speech, insults, restrictions, strict budgets and neglect.

On the same vein, caregivers in R_3 regard migrant parents as dodgers of parental responsibilities, abusive of caregivers and culprits who burden other people with the upkeep of their children. Migrant parents view caregivers as people who are after economic benefits of care and not care itself. In R_4, migrant parents suspect teachers to be poor, jealous and lacking understanding. On the other hand, teachers view migrant parents as uncaring, unloving and negligent with misdirected priorities.

R_5 is characterised by the distance between teachers and caregivers as well as shifting of responsibility of caring for migrant children. Teachers feel that responsibility of caregiving is shouldered on caregivers by migrant parents. Caregivers think their duty end at home while teachers think their role ends in the classroom. Caregivers and migrant parents think of teachers not only as providers of classroom knowledge but also as stewards, leaders, guardians, tutors and caregivers. R_6 is characterised by seemingly cordial relations between migrant children and their parents with few cases of misunderstandings which are quickly shifted to caregivers, teachers or school authorities. The tainted negative relations and the system of blame game create an environment which is not favourable for the social and academic life of migrants' children.

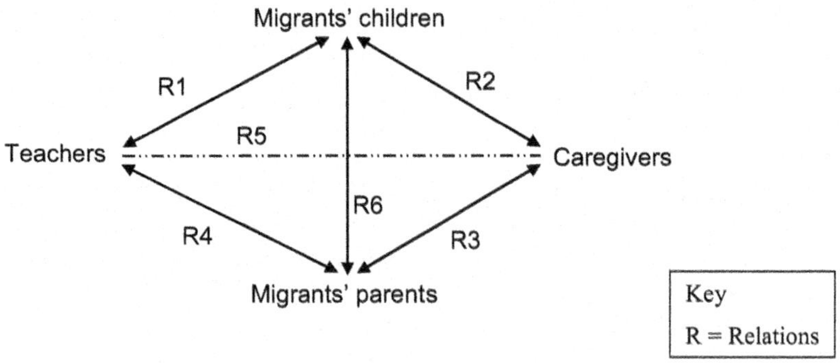

Fig. 7.1 Strained and suspicious relations

Conclusion

Teachers and caregivers do not seem to view migrants' children as having serious problems. To them it is migrant parents and their children who are burdensome and problematic. To teachers and caregivers, it is they who have negative experiences of caring for migrants' children. However, they acknowledge that lack of parental guidance, care and love are causes of all the problems they experience with migrants' children. It is also a fact that migrants' children experience neglect by migrant parents, teachers and caregivers. They also experience hate, mockery, abject poverty, abandonment and psychological problems that manifest in isolation, boredom, rebellious behaviour and hardened feelings. They live on the fringes of the society, school; caregivers' homes, migrant parents' lives experiencing marginality of the worst kind. They are children belonging to no one.

References

Baker C et al (2009) The impact of migration of children in the Caribbean Available from http://www.iunicef.org

Cortes R (2007) Children and women left behind in labour-sending countries: an appraisal of social risks. UNICEF, New York

Department of Education, Employment and Workplace Relations (2013) Family-school partnership framework: a guide for schools and families. Available from www.family school.org/family-school-partnerships-framework.pdf. Accessed 28 Apr 2020

FAWE (2006) Gender responsive pedagogy: effective schools and quality improvement. FAWE Draft Document, Libreville

Filippa OM (2011) Zimbabwe adolescents experience of their parents' absence due to diaspora. UNISA. Available from uir.unisa.ac.za. Accessed 12 June 2014

Karakus M, Savas AC (2012) The effects of parental involvement, trust in parents, trust in students and pupil control ideology on conflict management strategies of early childhood teachers. pp 2977–2985

Nakpodia ED (2012) Teachers' responsibility in loco parentis in secondary schools in Abraka. Metropolis, Delta State, Nigeria. Global J Human Soc Sci 12(11) Available from http://global-journals.org. Accessed 22 Aug 2014

Osthuisen IJ, (1992) The loco parentis role of the teacher: A relationship perspective. Koers 57(1): 121–134

Thamarasseri I (2017) Gender, peace and education. A.P.H. Pub Corporation, Cambridge

Von Roekel D (2008) Parent, family and community involvement in education. NEA policy brief. Available fromwww.nea.org/parental involvement.pdf. Accessed 28 Apr 2020

Wei Lu G (2011) Left-behind children in rural China: research based on the use of qualitative methods in inner Mongolia. University of New York, New York

Worley V (2003) The teachers' place in the moral equation: in loco-parentis. Oklahoma State University, Still Water

Yawkey TD, Bakawa LJ (1974) The child care professional/parent/child: an emerging triad. University of Winsconsin, Winsconsin

Chapter 8
The Impact of Migration on the Family and Children's Rights: Gendered Perspective

Debates on Children's Rights and Migratory Processes

Migration has moved up the ladder in the policy agendas but its implications on children have received little attention. Children are affected by migration whether left behind, born abroad, migrating alone, left by one or both parents. Impact of migration on children and adolescents should be seen in the broader context of poverty, vulnerability, resilience, gender relations and children's rights (UNICEF 2007).

States have an obligation to respect provisions of convention in the policies and actions towards each and every child within its jurisdiction. The CRC protects each and every child regardless of nations and immigration status (UNICEF 2007). As a result, on 28 September 2012, the committee on the Rights of the child devoted a whole day for the general discussion on the Rights of children in the context of international migration to raise awareness about children's rights. They provided inputs for state policy programmes and guidance to relevant actors on the respect, promotion and fulfilment of the rights of children in situations of international migration. Protection gaps in origin, transit and destination countries in the international framework on the rights of children were addressed. The concerns were human rights challenges affecting children such as views on the rights of children to promote collaboration between actors dealing with children's rights (Committee on the Rights of the Children 2012). The other concern was the universal treatment of children affected by migration despite their different contexts and categories. The categories of children affected by migration are refugees, trafficked, smuggled, asylum seekers, left-behind, unaccompanied, returned and the regular migrants.

Universal treatment of the above children indirectly leads to exclusion of some categories. All children are to be treated as children and their rights guaranteed regardless of their migration status or that of parents. Some children do not have access to civil, economic and social rights, for example, birth registration making children stateless. This further restricts their access to education, health care, housing, social and economic protection. Pregnant mothers have no access to maternity

W. Matsa, *Marginality, Migration and Education*,
https://doi.org/10.1007/978-3-030-60873-6_8

services thereby disadvantaging unborn migrants' children. Most of the migrants' children also lack documentation in both origin and destination countries. Left-behind children also lack access to preventive care and have no legal provisions for state-funded emergencies thereby excluding them from social protection measures (UNICEF 2012).

The restrictive migration policies place children at risks and situations of vulnerability. The migration status of children is always linked to that of parents. If parents are illegal, children are treated as illegal being denied every right that they should be entitled to. Left-behind children are also not known and are not provided for by their states whether parent migration is legal or illegal (UNICEF 2012). Children are subjects, in their own right and should actively construct their own lives. Unfortunately, they are absent on debates on ethics of migration despite that they make up 50% of displaced people (UNHCR in Josefsson 2014). Children are viewed as appendages of parents and caregivers. They have been conceptualised in a narrow traditional view of being not yet adults bearing no rights of their own.

Children are invisible as moral subjects or if mentioned are viewed as vulnerable subordinate part of the family. They are not seen as active participants in migration processes but unfinished adults rather than being in themselves. Traditions assume that the interest of the child is consistent with that of the family or parents. Parents, the family, or the state has legitimacy over the child and can convey the interests of the child. Migration policies and researches are rooted on this traditional dominant view of the child. Adults dominate migration discourses portraying children as passive and in need (Aristotle and Locke in Josefsson 2014). Perspectives of children are overshadowed by adult discourses and state policies. There is no focus on children as autonomous individuals with independent rights of their own.

The impacts of migration on left-behind children in Bulilima, Mangwe and Zimbabwe in general relates to the above dominant adult discourses. Disintegration of families, configuration of new forms of families, shifting care, residences and schools, have to do with adult decision-making processes. Decision-making does not take into account views of children. Migration processes are informed by the traditional view of children that they are unfinished adults, are not autonomous and their interests converge with those of parents. Even the situation of vulnerability is supposed to be understood and solved by parents, despite the CRCs and the state role in protecting the children. Decisions by parents expose children to violation of almost all rights of children and migration itself disrupts the family as a system.

The constitution of the Republic of Zimbabwe Amendment no 20 (2013), section 19 has the provision for enjoyment of family and parental care appropriate when children are removed from their biological family. Section 20 provides for access to shelter, basic nutrition, health care and social services. Children are to be protected from malnutrition, neglect or any form of abuse. They are to have access to appropriate education and training. Appropriate legislation is to be taken to protect children from malpractices to ensure that children are not placed at risks of well-being, education, physical, mental, health or spiritual, moral or social development.

Section 22 of the same constitution urge the state to take practical measures to promote free and compulsory basic education for children and that every citizen and permanent resident of Zimbabwe has the right to state funded and further education. Discrimination is prohibited by the constitution (section 74). Section 81 adds that there should be equal treatment of all children and that all children born in or out of Zimbabwe have right to citizenship and birth certificates. Despite these provisions, some rights of children are not observed by parents and communities in Zimbabwe. The government does not adhere to strict monitoring of measures provided for by the constitution of Zimbabwe.

Family Systems Theory

According to the family systems theory, individuals cannot be understood in isolation from one another but rather as part of the family and that the family is a system of interrelated parts (Bowen in Johnson 2010). The behaviour of individuals is a reflection of the family, so the family systems theory focuses on family behaviour rather than on individual behaviour (Fingerman and Bermman 2000). The family system has patterns of behaviour that influence its members. Relationships at family level are influenced by parents consciously or unconsciously.

There is a reciprocal and interdependent relationship that exists between the family and the broader social system. Social systems like migration, education, death, loss of a job and economic crisis affect the family system. A change in one part of the family system reverberates throughout the rest of the system (Bowen 1978; Johnson 2010). In the context of the family systems theory, even factors that appear to influence only one person, the migrant, have an impact on everyone, for example, migrants themselves, caregivers, teachers and the larger community. The family functions to socialise children for future performance and adaptation mechanisms of the family and the larger society.

At this juncture, it is crucial to operationalise the terms, family and family systems, and the term family referring to "…. two or more persons living together, related by blood or adoption" (Bluffton 2014). It is made of complex structures consisting of an interdependent group of individuals who have a shared sense of history, have emotional ties to one another and who devise strategies to meet family needs (Bowen 1978).

Families are systems because they are made up of interrelated elements (family members) that exhibit coherent behaviour, have regular interactions and are dependent on one another (Winter 2001). Interaction patterns are predictable in a system because they are based on rules and set boundaries. The family is a mechanical system made up of multiple parts. Non-function of one part impacts on the other parts (Bluffton 2014), thus the family system must be understood as a whole, and cannot be comprehended by examining individual parts in isolation from one another because it functions as a system.

The family is a basic social unit of all cultures, and through time, families have represented the most significant institution for nurturing, caring for and socialising children. The child's quality of life is directly influenced by the kind of care, support, stimulation and education he or she receives from family members in the home (Bluffton 2014). The family unit forms the foundation for secure membership in other groups throughout life and influences performance at every level. Parent and child ties in adulthood depend on ties in childhood (Fingerman and Bermman 2000). Migration affects the function of the family system as a nurturing and socialising institution.

The family system has sub-systems which include parents, children, grandparents and other relatives. The family also has well-defined strategies to achieve family tasks. The first-order task of the family is to socialise family members into attitudes, beliefs, values and emotional aspects that help families to interact with others (Bowen 1978). Changes in parental sub-systems as a result of migration may lead to changes in experiences of children of migrants, necessitating changes in caretakers and other parental sub-systems.

Families are also responsible for providing basic necessities like food, shelter and education for the sustainability of the family and all these fall under second-order tasks. It is also the role of the family to be responsible for managing the emotional climate of the family in a way that promotes psychological well-being of its members (Gharehbaghy 2011). The family, with or without parents, is expected to perform the above roles.

Interdependence or reciprocal relationship also exists between family sub-systems themselves. Changes in composition of the family to one-parent, parentless, child-headed or grandparent-headed households, among others, may lead to changes in children's autonomy or parent's behaviour. Guardian-parenting styles may also lead to changes in interactional patterns and psychological being of children in the family. Sub-systems and members are interdependent. Family systems should establish patterns of interaction, coping strategies of adaptation and management of family stability in the face of inevitable change like migration.

Families change (morphogenesis) and cope to maintain stability (morphostasis) of the family. Family systems should adapt to changes from both within and outside the family (Bertalanffy 1975 in Bowen 1978). Migration as a change may result from external or internal factors. It leads to economic, social and psychological changes that occur within the family. For this reason, the family needs adaptation strategies to manage the demands for change that occur within and outside the family system.

The family is expected to manage children's educational experiences resulting from social, economic and psychological stress as well as to establish coping or adaptation mechanisms for responding to and managing changes by migrants themselves, left-behind guardians and children themselves. These, according to Gharehbaghy (2011) are second-order tasks of families, where families need to adjust periodically in response to new developments that result in stress, isolation and lack of material, emotional and social.

The Government of Zimbabwe National Care Policy discourages institutional care and it is believed that children should be brought up in their own cultural contexts. The vulnerable children are to remain with the extended family. It is only in extreme cases where children who are orphans without any extended family can be taken to care institutions. As a result, many children with absent parents in Zimbabwe are in the hands of foster parents (Powell et al. 2004).

Industrial capitalism and modernisation caused the fragmentation of families and migrant labour has further intensified the disintegration of the extended family in Zimbabwe. Fostering is problematic in an urbanising society with the extended traditional families changing to nuclear families (Madziva 2010). Child-fostering is not formalised, resulting in children moving from one relative to another or even to people who are not relatives.

In the urban suburbs of Harare, Zimbabwe, when parents migrate, they surrender children to all sorts of people, ranging from grandparents, sisters, brothers, friends, abusive spouses and employees. Children left with foster parents pay the price of migration. The idea of leaving children behind is a "horrific trend" though thousands of children continue to be left in the care of family friends and relatives (UNICEF 2007). Makoni, a director of a non-governmental organisation, Girl Child Network (GCN), cited in UNICEF (2007), related an incident in which a girl left in the care of a foster mother in Harare was raped by the foster mother's boyfriend when her mother migrated to England. The girl's mother preferred a friend instead of her mother and sisters because the friend had a good job and was living in an affluent suburb. The friend who became the foster mother worked for a non-governmental organisation and often went out of the country. The child, a girl, became exposed to sexual abuse.

Ten cases of reported abuses in Zimbabwe concern children whose parents have left the country. Another 17-year-old girl was abused in an attempt to get a visa and passport to England through parents' friend. When the girl arrived in Malawi, there was no accommodation for her. She was taken to a brothel where she was treated as a sex worker and she was to sleep with affluent people to secure fraudulent travel documents. The trafficker took the money from her but she did not get the travel documents (UNICEF 2007). Although the experiences outlined here concern mostly children in urban areas, it also happens to rural children of Bulilima and Mangwe as far as shifting families and care is concerned. Although both males and females are affected, girls are affected to a greater extent by the burden associated with shifting of care and violation of rights to protection, reproductive health and sexuality, survival, development and health. As a result of culture, rights to identity and descent affected boys to a greater extent. A boy-child belongs to the father, should grow up with their fathers' people or unite with them for the survival of the lineage. A boy who does not know, or has no link with his paternal relative is looked down upon and mocked by general people and his maternal relatives.

Most significantly, parental migration in Zimbabwe in particular Bulilima and Mangwe interrupts the family role as an institution for socialisation, caring and nurturing. The parents are the significant others. The family no longer form the foundation for secure membership in other groups. Family disturbances tend to lead to disruptions in education. Although the family and the schools are two separate systems, they are interdependent. What happens in one institution affects the other. Changes in the family system, as a result of migration in turn cause changes in educational relations between learners and teachers, children and migrant parents, migrant parents and teachers as well as relations of caregivers and children.

New Forms of Families Elsewhere and Zimbabwe

As a result of migration in South African province of Mpumalanga, family members become dispersed leading to residential separation and high mobility of children. Translocal residential arrangements and scattering of children across multiple households was also common in Kwazulu. Families of migrant parents sometimes had members living in different residents. Children of the same family ended up not receiving the same upbringing (Bennett et al. 2014). Dispersal of family members creates new forms of family and residence.

Migration of mothers in Zimbabwe displaces and disrupts not only the family structure but care relationships. The extended family and kinship networks are shrinking and migrant parents rely on the extended family carers and friends. On the other hand, males feel families are incomplete without the mother. Where both parents migrate, children either are left alone under the care of other older siblings in most cases boys for security reasons. The eldest girl plays the motherly role of the nuclear family. Sometimes children are cared for by members of the extended family. Members of the extended family prefer boys if they are farmers to benefit from hard labour. Generally, relatives prefer girls for household chores and girls are easier to control than boys. When parents migrate, life changes for the better financially, although children suffer from emotional stress. Comfort from the extended family and caregivers is not enough.

Madzivadondo (2012) has it that 3–4 million men and women have left Zimbabwe to the diaspora. The family as the epitome of development is disrupted. Zimbabwean exodus to the diaspora leaves a scar in the family as an institution. In some families, parents and children are left alone while, in some, both parents and children take different directions. Where the father migrates, the family is more cohesive since the mother remains to care for the children.

Labour migration in Bulilima and Mangwe results in disintegration of the nuclear family giving rise to new forms of family and parenting styles. These include care by non-blood relatives, for example, teachers, neighbours and employed caregivers among others. New forms of families include child-headed, elderly-headed, child renters and worker-headed families. Parental migration has also resulted in emergence of new forms of child-headed families as opposed to the known common type where siblings left by parents live alone in their home without an adult to take care of them. One new type is the child employer (worker)-headed, where a male head or female housekeeper sometimes younger than the legal age of majority (18 years) takes care of migrants' children, their home and the property.

The definition of a household has also broadened to include children from different households of the extended family who do not share residence, but decide to share sleeping accommodation for security reasons. They live in different homes, cook and eat separately but come together as children in the evening to sleep together in one home for safety.

Some children stay alone in their homes belonging to parents (child-headed families) where the older sibling takes care of younger ones, or a child is alone taking care of him or herself. This is common with boys. There is under-protection and lack of supervision by any adult. In Romania, these children are referred to as "child alone" (Botezat and Pfeiffer 2014), and in Trinidad and Tobago it is called self-guardianship where 14- to 18-year-olds assume surrogate parental roles to look after themselves (Baker et al. 2009). In Zimbabwe, "child alone" is common to boys than girls.

Another group of children stay alone in rented houses (child renters) or stay as a group renting a full house. These can be 2–3 in a room meaning that a 5 roomed house can have as many as 10–15 depending on the size of the rooms or the expectations of the property owner. The house can accommodate both boys and girls though girls or boys have their separate rooms. These children lack adult supervision or care in these child-headed illegal boarding houses. This is a drastic shift from the traditional nuclear family or extended family to a new form of residence bringing children from different families to a single household when they have no relationship. It cannot be equated to institutionalised care because the houses are not institutions with institutional rules and set up. Without adult supervision, the living arrangement has temptations of fornication and co-habitation. Girls are more lightly to lose time playing "wifely" roles at the detriment of schooling. Girls also get pregnant and drop out of school.

These are similar to surrogate families in South East Asia where left-behind children are brought together in a family style environment (Wei Lu 2011). However, in South East Asia, there is a paid guardian who takes the role of a parent. As a result, the Far East Asia such surrogate caring actors are referred to as "dollar mommies" working in care stations or institutions meant to care for children with absent parents due to migration. Professional surrogate parents provide food, accommodation and supervision of study, among others. The surrogate parent is paid by the parent of the child and there is constant contact between the two. A contract is also made between the two. Surrogate families have their own advantages. The contract between the parent and the surrogate parent is legal, making the child more secure through constant guarding. In Zimbabwe, the illegal and uniformalised boarding houses lack all the above.

Other forms of families in Zimbabwe are male and female-headed, grandparents-headed and worker-headed families. Male-headed are those where females migrate leaving children with their fathers, uncles, grandfathers and male employees although this is not very common. There is also increasing number of de jure and de facto headed households where a woman spouse acts as a head in the absence of a male head who is traditionally known as the decision-maker (Kethusingile et al. 2004).

In Bulilima and Mangwe, migration disrupted conjugal nuclear families and gave rise to various forms of families. Family structures were redefined. Working diaspora families living apart and dual families emerged. Where one spouse is left in Zimbabwe by a migrating spouse the conjugal nuclear family is disturbed, the spouses live apart. There are instances where sons and daughters establish their nuclear families in the diaspora leaving their parents' families behind. In Zimbabwe, they are still part of the bigger families and when they are in Zimbabwe, they live together as a single family. Sometimes one spouse who is in the diaspora establishes a nuclear family in the diaspora when he or she left another nuclear family in Zimbabwe giving rise to dual families.

Migration also causes re-constituted families like child-headed families of nuclear kin composition of siblings, which is bringing children from different nuclear families of kins to constitute one family. The older siblings look after younger ones. Older girls take up the care role of mothers and boys take up the protective role of fathers. There is also rise of multigenerational families where great grandparents, grandparents, aunts, uncles and cousins are left as a single family for care of the elderly and children. Sometimes in this case, there is merging or blending of two or three families into one extended family. Migrants' children become part of these families. Sometimes ageing families are left alone taking care of migrants' children. Migrants' children are fostered or adopted into these families.

Sole families are also evident where teen male children or caretaker workers are left alone by parents working in the diaspora. Step families also exist though they are few. Step families exist when two people get married and blend two families from their previous marriages then one of the spouses migrant living children under the care of the step father, mother or relative. There are also patrilocal and matrilocal families with authority resting on fathers or mothers, respectively. Types of families were not quantified, as this was a qualitative study, but learners sampled and caregivers interviewed were part of these various types of families.

Shifting Families and New Forms of Care

The shifting of care in Zimbabwe creates a chain of care relationships in the caring trialogue of caregivers, migrant parents and left-behind children. Conflicts, finger-pointing, responsibility pushing and black-mailing arise. This results in increased child shifting, shifting of care and residence. Child-shifting is defined by Baker et al. (2009) as a movement of a child from one home or one guardian to another. It can involve also a movement to another location with the new caregiver causing also change of schools and schooling environment. Family members become dispersed and residential arrangements change. Families of migrant parents in Zimbabwe sometimes have members living in different residences with different caregivers. Children belonging to one family end up in care and residences different from what they are used to. The circumstances in which children live become dynamic and complex.

Some children share residence with paternal while others live with maternal kins. This prevents children from the same family from receiving the same upbringing. Some children live with well-to-do families while others live with poor families.

Breast-feeding children sometimes go with parents only to be sent back to grandparents or relatives when they reach pre-school years. Residential instability place children at risk and reduce their well-being. Boys shifted more than girls because they are difficult to discipline. They sometimes move from one relative to the other until they are old enough to live alone. Some boys prefer travelling long distances from their parents' homes than to live with relatives. This is because of their patriarchal nature of wanting to dominate, make decisions and demand freedom in all aspects of life. Caregivers do not accept such behaviour.

In Zimbabwe, Bulilima and Mangwe in particular, some children are born when parents are already in the diaspora and stay with parents up to 4 years. They are then sent back home to grandparents for pre-school (1 year) and primary school (7 years) giving a total of 8 years. If there is no nearby secondary school, the child shifts residence and care to a relative, foster parent or to rent accommodation. From form 1 to form 2, they can stay with a foster parent for proximity to school. If the rural day secondary school does not have upper sixth form, a child shifts to another school. This also results in shift of residences and care. For financially stable parents, usually in forms 3–6 or examination years, forms 4 and 6, children, in most cases girls are taken to a conventional or low-cost boarding for better academic environment. Up to form 6, a child can change four residences, caregivers and schools.

Migration and the Violation of Children's Rights in Zimbabwe

International migration and its impact are currently receiving unprecedented attention at policy level in both sending and receiving countries (UNICEF 2014). International migration impinges on the rights of migrating children and those left behind. The impacts are largely felt in areas of health, security and education of children. Children left behind by migrant parents are increasingly affected by migration, as such the impacts of it should be positioned within the context of the human rights framework that provides for the protection of all children (Abramovich et al. 2011).

Children and adolescents than adults are at greater risks of violations of their rights due to direct or indirect effects of migration. Globally, there are 83 million international migrants under the age of 20 years which is 15% of all international migrants though there are regional variations (UNICEF 2014). Migration policies and processes lack provisions for the protection of children's rights. Migration is viewed as a strategy to increase opportunities, improved standard of living and the welfare of the family through finances or goods. Negative effects of migration on children are ignored. The majority of children suffer some kind of constraints on their rights in their country of origin or destination. These range from poverty, violence, deprivation of rights to protection, education, health and the right to parental guidance for those left behind.

In Moldova, Child Rights Information Centre (CRIC) noted that international migration violated a number of rights for migrants' children. The violation of one right led to the violation of other rights. For example, first to be violated was the right of children to be with their parents' Article 9 on the CRC. This automatically violates Article 5, the right to parental guidance and the best interest of the child Article 3 was ignored. The study conducted in 2005 in three geographical communities of Moldova found that children between 10 and 18 years were vulnerable to drug abuse risks violating the right to protection from any harmful use or abuse or trade in drugs Article 33 (Gavriliuc 2006; UNICEF 2014).

Although material conditions of life improved in Moldova, children experienced unpleasant emotional stress, sadness and insecurity resulting in violation of the right to moral and emotional development which comes with parental responsibility and guidance as nurturers (UNICEF 2014). The departure of parents modify children's sphere of relations and quality of life and health. Parents leave when children are not yet prepared for independence, lack management of resources as well as self-management leading to unhealthy life styles. Through this, the right to health is violated (Ankut 2014). Children's opinions concerning departure of parents are rarely considered in the family decision-making. They deny children the right to participation in decision-making and the right to be heard, the right to express opinions and to share information (Article 12 and 13) (UNICEF 2014).

In Trinidad and Tobago academic performance changed and school efficiency decreased because of lack of parental support and encouragements. At local level, schools are not concerned with the situation of migrants' children because they are considered more privileged than vulnerable. This violated their right to quality education and non-discrimination of children on the grounds of race, religion, type of family or unfair treatment on any basis (Gavriliuc 2006).

Migration by parents leaving children behind violates the rights of children enshrined in the Convention on the Rights of a Child. Article 5 of the United Nations (UN) Convention on the Rights of the Child (CRC) (UNICEF 2014) provides for parental guidance and governments are expected to respect the rights and responsibilities of parents to guide and advise their children as they grow up. In Bulilima and Mangwe, Zimbabwe, children are denied that right. Governments should also assist families to fulfil their essential role as nurturers of children. There are no policies relating to migration of parents leaving children behind. No assistant given to vulnerable children left behind by migrating parents.

Article 9 discourages separation of children from parents unless it is in the best interest of the child (Article 3) which is of primary concern in the convention (UNICEF 2014). In Zimbabwe, children are not even consulted on migration of parents, shifting of care and schools. They are not part of decision-making on issues affecting adults. Adults do not consider the best interest of the child but in most cases what is affordable and convenient to them as parents.

Every child has a right to stay in contact with both parents unless if this might harm them. Some children have never seen their parents or last saw them when they were young. Some parents do not return, do not phone regularly or never phone. However, government's role on issues of vulnerability of children, especially those left behind by parents due to international migration, is not visible even in the

Charter. The rights of children in this area are neglected by both the parents and the governments.

Article 28 provides for the right to education. The convention places a high value on education encouraging young people to reach the highest level of education which they are capable of attaining. Governments are expected to deal with issues that infringe on the rights of children in this area (UNICEF 2014). Migration is one of the issues impinging on education and needs to be looked into. Intervention measures taken by governments to ensure that children left behind by migrants' children access primary and secondary education up to the level they are capable of are non-existent. As a result of absence of parental guidance, girls drop out of school to marry because of pregnancy. Boys engage in illicit behaviours, abuse of drugs and alcohol which is detrimental to schooling.

The CRC was adopted in 1989 by the UN and received near-universal ratification and went into force within months of adoption (Kaime 2009). Zimbabwe is one of the countries that ratified it. The consensus affirmed shared global recognition of the rights of the child, the need to promote and protect the rights of the child. The convention coincided with global concern for education that was reiterated at a series of international conferences on Education for All (EFA) in Jomtien, in Thailand, which made a call that primary education should be made accessible to all children (UNESCO 2008). Zimbabwe is one of the countries that implemented EFA with recognisable success strides.

The African Charter on the rights and welfare of the child was the first regional treaty on the rights of the child, and it was drawn from the CRC (Ankut 2014). The charter was adopted by the Organization of African Unity in 1990 and became operative in 1999. Zimbabwe recognises the African rights charter and works towards its fulfilment.

In April 2000, the World Education Forum convened in Dakar, Senegal, to reaffirm nations' commitment to education as a fundamental right and to the goals of EFA (UNESCO 2008, p. 8). The African Charter, just like the CRC lacks focus on children left behind by migration. The right to education and other rights of children are violated everyday of children's lives with severe consequences which extend beyond their childhood (Ankut 2014). Violations extend from home to school and the community. Failure to provide for education and protection of children often translates into violation of their other rights (Sward and Bruum 2007). Universal education was declared as a right for all children including those in difficult circumstances (Schapiro 2009). Children separated from parents due to international migration are in difficult circumstances because they experience double vulnerability. They are vulnerable both as children and as children lacking protection because of absent parents.

Length of stay, lack of frequent visits and absence of parents, has negative effects on the education of these children. Parents have no contact with their schools and know very little about their schooling. This is a result of the length of migration and the frequency of visits. Migrant parents visit during Easter and Christmas holidays, when schools are closed. Their migrant parents are not in touch with their teachers and have no time for consultations or checking of books. Those who come during the term arrive on Friday evenings if they are from Botswana or Saturday if they are from South Africa and return on Sunday mornings. This happens when there are funerals or serious illnesses. Migrant parents rely on reports by caregivers who sometimes do not attend school meetings or consultations. The visits are short, busy and coincide with school vacations and weekends. Children are denied the right to parental guidance (Article 5 and 18) and educational assistance (Article 28 and 29).

Taking into account the fact that some children are born when parents are already migrants and that the majority have a length of migration of more than seven years, children of migrants can complete primary or secondary education or both without their parents practically participating in their education. Migrant parents take intervals of one, two, four or seven years while others never return. Children lack exemplary life from parents. Boys lack parental while girls lack maternal guidance. For those migrant parents who rarely return, communicate or never return, children are subjected to parental neglect, Article 5 of the UN CRC (UNICEF 2014).

Although some have frequent communication with their migrant parents, this is not very useful in their education. Their parents do not practically take an active role in their schooling but depend on reports from their children, caregivers and teachers. Sometimes they depend on hearsay from children or caregivers only, getting a one-sided story. Sometimes children lie to parents especially boys.

Of interest to note, is that deprivation also translates into violation and lack of respect of many other rights through deprivation of parental guidance. High regard for education by Article 28 is not observed. Young people are not given opportunities to reach the highest levels of education they are capable of. Children of migrants are unprotected, insecure and vulnerable to harmful use and abuse of drugs, alcohol and unhealthy habits leading to a poor quality of life (Article 4, 6 and 33). Protection rights are not fulfilled for child survival, development and health living. Migrants' children lack medication, medical aid cards (Article 24) on health and health services.

They also do not have birth certificates thereby denying them the right to access health (Article 7 and 6), education (Article 28) and the right to belong to a family (Article 8). Children are to be legally registered to be officially recognised by the government and to know their names, parents, nationality, descent for belonging in order to identify who they are in their families and nations. Without registration, they become stateless. Issues of patriarchal descent affect boys to a greater extent. Children's rights to health, protection from harmful substances, right to belonging, care and parenting are violated severely. The violations are unrecognised or under-reported, leading to creation of barriers to the development and well-being of children.

Parents are to take responsibility to guide and advise their children as they grow up (Article 5 of the UN CRC on for parental guidance). Parental migration in Bulilima and Mangwe violates this right. The government should take responsibility to assist families in fulfilling their essential roles as nurturers and carers of children. Article 9 discourages separation of children from parents but in Bulilima and Mangwe, the separation occurs over the whole period of school life or even over a lifetime. Children according to the convention are to live or to be in contact with parents unless it is harmful for them to do so. Article 10 provides for re-unification of families. Family members are to move between countries so that parents can stay in contact or get back together as a family. Some migrant children expressed wish to migrate to re-unite with parents but this is not possible for some families in Zimbabwe because they migrate illegally. The government has not done anything in this direction to ensure migrants stay in contact with their children. This consequently leads to violation of Article 12 of respect of views of the child to say what they think about the situation of migration that affects them. Adults do not listen to children but deny them freedom of expression (Article 13).

In Zimbabwe, migrants' children are also deprived of family environment where children are looked after by their own parents (Article 20). Migrants' children can be looked after by caregivers until they become adults themselves. If they also become migrants, they also give their children to caregivers. There is a rise of a phenomenon of parents who have never experienced parenting children of their own. The definition of parenting for migrants' children changes to guardianship or caregiving. Some migrants have no experience of parenting and some migrants' children have never been parented. This will give rise to generations of people lacking parenting skills.

Article 28 places high value on education, encouraging young people to reach the highest level of education they are capable of. No provisions are made for parents migrating with children or those leaving children behind. The home and school environments do not provide for the reaching of the highest level of education because of challenges of lack of material resources for education, educational and moral support, influence of culture of migration, parental discipline and generally lack of conducive environment for learning. The goals of education to develop the child's personality, talents, respect for human rights, respect of culture, parents, teachers and other people cannot be achieved (Article 29). Although governments agreed to review laws and policies relating to children accessing social services and funding for children to meet the minimum standards, the government has not categorised migrants' children as vulnerable and there are no laws or policies relating to the situation of migrants' children.

Article 2 provides for non-discrimination. No child should be treated unfairly on any basis of their race, religion, type of family, where they live, language spoken, gender, disability and poverty among others. Migrants' children are treated unfairly by teachers and school authorities who prefer children of non-migrant parents.

Children renting accommodation in business centres and those in illegal boarding houses and bush boarding experience sexual and violence despite Article 19 that provides for protection from all forms, mistreatment, emotional abuse and neglect. There is no review of care in these living arrangements as provided by Article 27 meant to ensure adequate standard of living. Some lived in poor dilapidated houses or huts with no security, drinking untreated dam or well waters. The premises are noisy because of bars and night clubs. Such living conditions lack social security, (Article 26) and do not meet the physical and mental needs of learners thereby violating Article 24 on health.

Article 21 provides for the right to adoption or foster care considering what is best for the child. For migrants' children, it is not formalised foster care. There are no rules binding the foster parent on how to care for the foster child and no contract signed with parents. There is no consideration of whether the foster parents are the best choice for the children and no one oversees the upkeep of migrants' children during the absence of migrant parents.

Conclusions

Amid all these problems associated with new forms of family, shifting care and violation of rights, children remain voiceless. There is absence of child rights-based perspectives in migratory policies and legislation in Zimbabwe and this is a notable concern. Child perspective in educational and social problems is also lacking. Protection policies by government and non-governmental organisations in relation to migrants' children are non-existent. In governments and development agencies migratory policies and regulations, migrants' children have not been conceptualised and categorised as vulnerable, marginalised and at risk yet their education, rights and life in general testifies of marginalisation. There is still scarcity of reliable data at country level on incidence and magnitude of the impact of international migration on adults and children left behind (UNICEF 2014). Focus has largely been on the education of migrants' children at their destination. More research is needed to make children's experiences more visible in migration debates and policies. The continued absence of close family members like parents disrupts child development and schooling. Policy is needed to tackle negative effects of migration for the children left behind.

References

Abramovich V, Cernades PC, Morclachetti A (2011) The rights of children, youth and women in the context of migration: conceptual basis and principles for effective policies with a human rights and gender based approach. UNICEF, New York

Ankut PY (2014) The African charter on the rights and welfare of the child: linking principles with practice. Available from http://www.hurisa.org.za. Accessed 15 Jan 2014

Baker C et al (2009) The impact of migration on children in the Caribbean. Available from http://www.unicef.org. Accessed 23 Jan 2014

Bennett R et al (2014) An approach to measuring dispersed families with a particular focus on children left-behind by migrant parents: findings from rural South Africa. University of Kwazulu Natal, Durban

Bluffton C (2014) Family systems theory. Available from http://www.bluffton.edu. Accessed 29 Dec 2013

Botezat A, Pfeiffer F (2014) The impact of parents' migration on the well-being of children left behind: initial evidence from Romania. Discussion Paper No. 6225. Centre for European Economic Research, Mannheim

Bowen JM (1978) Family systems theory and practice: illustration and critique. Available from http://www.thebowencentre.org. Accessed 29 Dec 2013
Committee on the Rights of Children (2012) The rights of children in the context of international migration report 2012. Available from https://www.ohchr.org/document. Accessed 28 Apr 2020
Fingerman KL, Bermman E (2000) Applications of family systems theory to the study of adulthood in ageing and human development. Baywood, Washington, DC
Gavriliuc C (2006) The situation of children left behind by migrating parents. GFMD, Available from http://www.google.co.zw. Accessed 27 Dec 2013
Gharehbaghy F (2011) An investigation into Bowen family systems theory in an Iranian sample. Iran J Psychol Behav Sci 5(1):56–63
Government Printers (2013) The constitution of the Republic of Zimbabwe Amendment no 2. Government Printers, Harare
Johnson S (2010) Family systems theory. Available from www.familytherapy.vt.edu. Accessed 10 May 2013
Josefsson J (2014) Children as moral subjects in ethics of migration. In: Proceedings from the 49th societas ethica Annual conference 23–26 August 2012. Journal revista ecumnica sibiu, vol IV: 3, pp. 148–157
Kaime T (2009) The African charter on the rights and welfare of the child: a socio-legal perspective. Pulp, Cape Town
Kethusingile BM, Kwaramba A, Lopi B (2004) Beyond inequalities: women in Southern Africa. SARDC, Harare
Madziva R (2010) “A living death”: Zimbabwe migrants in the UK who are forced apart from their children. University of Nottingham, Nottingham
Madzivadondo T (2012) Impact of spouses in the diaspora on the family as an agent of socio-economic development in Zimbabwe. J Soc Sci Res 1(6):196–199
Powell G, Chinake T, Mudzinge O, Maambira W, Mukutiri S (2004) Children in residential care: the Zimbabwean experience. UNICEF. Available from http://www.crin.org. Accessed 12 June 2014
Schapiro K (2009) Migrants and educational outcomes of children. In: Human development research paper 2009/57. Available from http://hdr.undp.org. Accessed 10 June 2013
Sward S, Bruum L (eds) (2007) Focus on children in Migration: From a European Research and Method Perspective. Warsaw: Save the children
UNICEF (2007) The impact of international migration: Children left-behind in selected countries of L. America and the Caribbean. UNICEF. Available from www.unicef.org. Accessed 17 Jan 2014
UNICEF (2008) The effects of parental deprivation on the development of the child left behind by Moldovan migrants. UNDP/CRIC, New York
UNICEF (2012) “Access to civil, economic and social rights for children in the context of irregular migration” UNICEF Written submission on 2012 Day of general discussion of the committee on the rights of the child. Available from https://www.unicef.org/socialpolicy. Accessed 28 Apr 2020
UNICEF (2014) A summary of UN CRC. Available from Childrenandyouthprogramme.info/pdf. uncrc-summary-version. Accessed 28 Jan 2014
Wei Lu G (2011) Left behind children in rural China: Research based on the use of qualitative methods in inner Mongolia. University of New York, New York
Winter CM (2001) Family Systems Theory.web.pdx.edu/cbcm/CF5410U/FamilySystemsTheory.pdf

Chapter 9
"Silent Wars" Between Migrants' Children and Teachers, Teachers and Migrant Parents: Gendered "War"

Teacher–Parent Partnership in the Schooling of Children: The Impact of Teacher–Parent and Teacher–Student Conflicts on the Educational Process

Achieving cooperation and collaboration between teachers and parents is a challenge (Lasater 2016). Teacher–parent relationship determines teacher–student relationship. Family–school partnership strengthens the connection between the family and the school as well as learners and other community members. It is an antidote against conflicts (Department of Education Employment and Workplace Relations 2013). For a positive school climate and success of education to exist, there should be a harmonious relationship and cooperation between schools and homes (Grazin and Brigita 2017).

Conflicts between teachers and parents affect learners negatively. Teachers and parents have to keep conflicts private from students or resolve them as soon as possible to re-establish positive genuine relationships. Unresolved teacher–parent conflicts impact on the family–school partnership. The conflicts cascade to teacher–student conflicts. When teachers are in conflict with parents, they should take note of what they say about parents in the presence of the child. Both the teacher and the parent should be careful with verbal and non-verbal behaviour towards one another. Teachers should be trained on how to handle problem of parents who approach and attack teachers. Barriers to resolving conflicts have to be done with distrust between educators, parents, and perceptions that differ, as well as different perspectives about students (Lasater 2016). The conflicts that arise between migrants' children and teachers in Bulilima and Mangwe are closely connected to the relations of teachers and migrant parents. The perception and perspectives that teachers have about migrant parents are sources of conflict with migrants' children. There is lack of partnership and a lot of distrust. Unresolved conflicts worsen the bad relations.

W. Matsa, *Marginality, Migration and Education*,
https://doi.org/10.1007/978-3-030-60873-6_9

There is little data on teacher–student conflicts though they do exist. Teacher–student conflicts damage and disrupt teacher–student cooperation in the educational process. Conflict between teachers and students is usually associated with student behaviour. In reaction, then teachers try to use conflict management strategies that fuel and worsen the conflict, for example, dominating and using forceful means by imposing power they possess as teachers (Grazin and Brigita 2017).

Teachers may try to manage conflicts through withdrawal, verbal abuse, avoidance, compromise, inaction, ignoring or sending away. These worsen than resolve conflicts. The conflicts may remain unresolved and an unresolved misunderstanding generates more conflicts. More conflicts mean negative environment for learning. Teachers may decide to give unjust punishments or psychological torture, for example, insulting, humiliating, shouting and threatening students. Students feel humiliated, shamed, offended, annoyed and under-estimated. Such disciplinary measures create salient conflicts.

Students can react by arguing, blaming, lying, threatening or even scolding the teacher. They may also refuse to perform tasks given by the teacher including classroom tasks like solving a problem on the board or doing homework. Such reactions deepen the feud between teachers and students. Children with frequent and unresolved conflicts with teachers exhibit psychological problems such as aggression, rudeness, anxiety and low self-esteem. Students then lose motivation to learn and start skipping school (Grazin and Brigita 2017). In such situations, the teacher becomes less respected and students develop a negative attitude towards the teacher in question. The other party's behaviour triggers the other's response.

Interpretation of situations depends on beliefs or an attitude towards the other party. It also depends on the emotions that the parties carry in relation to the conflict and towards each other. Emotions affect interpretation and resolution of conflicts. Conflicts between teachers and parents, teachers and students should be avoided at all costs for they damage the educational process (Grazin and Brigita 2017). Attempts should be made to sustain partnerships and cooperation despite conflicts. Conflicts need quick resolutions before they escalate to attitudinal problems which are more damaging. In Bulilima and Mangwe, conflicts between migrant parents and teachers, teachers and migrants' children have deepened to attitudinal manifestations based on beliefs they hold about each other.

In Ecuador, migrants' children were viewed negatively. Some were associated with poverty and others with rapid social mobility. In Moldova, part of the teaching staff had mixed attitudes and feelings towards children without parental care. There was a combination of sympathy and envy of remittances. Teachers had meagre salaries, hence asked money from these children secretly as a reward for attention and high grades. Those with access to money were leaders in the class. These children were not judged according to academic performance or behaviour but according to the labels and attitudes of teachers and school authorities. They were labelled as arrogant, aggressive, rude, spoilt and extravagant (Cortes 2007). Most children in schools in China, teachers considered them hard to educate hence gave up disciplining them (Wei Lu 2011).

Save the Children (2014) highlights that teachers in Sri Lanka were expected to provide protected space for children without adequate parental care. Teachers were viewed as "surrogate" parents. Gavriliuc (2006) noted, however, that the same teachers considered these children privileged than vulnerable. Non-migrant children and teachers in schools labelled migrants' children as "barrel" children because of parcel boxes sent from abroad.

In Moldova, there were concealed conflicts between teachers and children of migrants (Cortes 2007). Their parents were viewed as "social climbers" and so were the children. Lack of cash and other material resources for academic purposes was not viewed by teachers and caregivers as genuine but as lack of care and seriousness about schooling. In Moldova, teachers were envious of migrant children and strongly felt that the wealth parents sent to their children made them do wrong things. Parents were reported to spoil their children by sending them those consumables like expensive cell phones, recorders, satellite antennas which distracted them from schooling. These were confiscated or withdrawn from learners by teachers. In line with children's rights, the government and parents allowed them to have the gadgets. Between 1993 and 2000, increase in delinquency crime rate positively correlated with the number of left-behind children who accounted for nearly 60% of offenders (UNICEF 2007).

In Nigeria, it was discovered that although the teachers were viewed as having full rights to mould children's moral character, assist in mental and physical development, their hands were tied (Nakpodia 2012). Teachers were not responsible for the behaviour, safety or protection of children outside the school after hours. Parents were the ones who were responsible for attendance and punctuality. Teachers did not supervise students during lunch hours or in sports outside school hours. The school itself had no regard to physical, economic and psychological well-being of students. This, however, limited the role of teachers to instruction and tutoring. It is worse for migrants' children who lack that care and protection outside school.

In Zimbabwe, teachers as loco-parentis are expected not only to be providers of academic knowledge, but also to be counsellors, disciplinarians and to give direction to the lives of these children. It is unfortunate that teachers have given up on these children and paint a gloomy picture of them. Teachers label migrants' children with negative stereotypes of indiscipline, deviants and notions of superiority. Teachers view them as concentrating in fashion, promiscuity, caring less about education. Teachers also felt that these children have a lot of money at their disposal and they boasted that they can pay teachers' salaries. This creates friction between them and teachers resort to leaving them alone to do as they wish (Kufakurinani 2014).

Kufakurinani (2014) further mentioned that teachers reported that migrants' children have a lot of freedom, independence and misuse money. They also have expensive and latest technologies like iPods sent by parents as compensatory love which does more harm than good. During lessons, these children boast with gadgets instead of using them for educational benefits. Parents maintain affection and control of their children through material things. In boarding schools, teachers cite that migrants' children throw mini parties in hostels and day-scholars hold big parties with friends using money and remittances from abroad. Attempt by teachers to counsel these children results in hurtful words directed to teachers, reminding teachers of their poverty. This behaviour causes friction between teachers and migrants' children. As a result, teachers do very little to control the behaviour of these children. Children of migrants in Bulilima and Mangwe also look up to teachers as providers of knowledge and support in their challenges but the attitudes they hold towards each other are a barrier to learning.

Behavioural and psychological problems make migrants children to be easily annoyed, sensitive and hostile to teachers and caregivers (Tufis 2007). They get depressed and are full of feelings of dislike. They battle with feelings of hatred and do not have good relations with teachers, caregivers and to some even their parents. Filippa (2011) discovered that the adolescents with diaspora parents in Harare, Zimbabwe, developed defence mechanisms that were negative. These include code of silence, denial, sublimation, displacement with substitute activities like isolation, chatting on phones, watching television all these being signs of bottling up one's feelings. The adolescents show irritability when they speak about teachers and caregivers. There is a strong link between absent parents and stressful life among these children.

Children between 15 and 17 years exhibited rebellious attitudes towards teachers and caregivers. They were also stubborn, disobedient and difficult to look after (Save the Children 2014). In Trinidad and Tobago, Baker et al. (2009) discovered that children had high levels of depression and were suicidal. In some cases, children are attracted to material possessions at the expense of moral values like love and respect for parents and other adults. These are the behavioural characteristics that make migrants children to have problems with teachers in schools yet teachers ignore their stewardship role.

The role of the teacher can be tracked back to the Greek pedagogues, a teacher who served as a steward, leader, guardian, protector, tutor and caregiver (Worley 2003, p. 281). This was a slave assigned to a boy everywhere outside the home, assist in lessons and discipline. The term later evolved to mean attendance on children. Pedagogues like parents, maintained, protected and educated children in care. Teachers like parents are to protect children from conditions harmful to learning. If teachers fail to provide reasonable care, they are liable to charge. Teachers are regarded as loco-parentis and are expected to understand material and psychological needs of children they teach. Loco-parentis means that they stand "in place of parents" (Hall and Mannis 2001, p. 117). However, Cortes (2007) noted that migration in Jamaica fostered certain forms of rejection by teachers, caregivers and the community at large. Non-migrant parents, caregivers, teachers and school authorities held unfounded negative opinions about migrant children which affected their performance.

Attitudes of Teachers Towards Migrant Parents in Bulilima and Mangwe

Teachers are to be viewed with respect in the community. The silent wars impact negatively on the integrity of teachers as academics, professional and "loco-parentis". Teachers' sour relations with migrants and their children impact negatively on the social status of the teacher, the teachers' authority in schools and the education system as a whole.

Relations between migrant parents and teachers in Bulilima and Mangwe are strained and teachers vent their anger on children of migrants. They insult migrant parents through children and hold unfounded negative opinions about migrants' children. Sometimes there are direct attacks on migrant parents citing that migrant parents are foolish because they buy flashy cars when they do not have houses and that migrants' children want to join their fathers in the washing of dishes in South Africa.

Teachers turn down requests or payment plans from migrant parents, expecting fees to be paid once in full arguing that migrant parents have the money because they make a lot of noise from their diesel and car battery charged radios during the December and Easter holidays. Sometimes teachers take advantage of the situation to take revenge on migrant parents, vent feelings of hate or competition with migrants. Teachers seem to be silently competing with migrant parents. They insult both children and migrant parents reminding them that there is nothing to boast about when migrant parents cross boarders to wash dishes for women and other men in the diaspora.

Teachers think *njivas* are rude, boastful and extravagant. They think njivas are envious of them for their academic and professional achievements. Male njivas also have a competitive advantage over teachers when it comes to women. Male njivas and male teachers clash over girlfriends in the community.

Attitude of Migrant Parents Towards Teachers

Njivas suspect teachers of having adulterous relationships with their left-behind wives, girlfriends and fiancés. Male teachers are also thought to be having love relationships with school leavers and learners some of which are migrants' children. Njivas also propose female teachers vying into the territory of male teachers. With the economic situation in Zimbabwe, some female teachers prefer *njivas* than teachers for their economic advantage of foreign currency. Of late, some of these *njivas* are well-educated and experienced professionals who left the country to the diaspora because of the economic and political situation.

Migrant parents criticise teachers for ignoring their role of loco-parentis and educational assistance. They think teachers have all the responsibility to ensure that learners do their school work, pass and behave well when teachers think that such is not their responsibility. Migrant parents feel teachers do not do enough and lack exemplary behaviours. They think teachers look down upon them assuming that all *njivas* lack education.

Teachers boast of their education belittling migrant parents because of the kind of jobs they do in the diaspora. This does not go down well with migrant parents. Some migrant parents view teachers as poor. This is exacerbated by the fact that teachers buy goods on instalments from the *njivas*. Some items are expensive for teachers to afford. Some teachers cannot afford cars but instead buy bicycles, second hand cell phones, iPads, radios, used car batteries and wrist watches. Other items which teachers buy from *njivas* are blankets and clothing of quality, for example, denim jeans and jackets. Such goods are unaffordable by teachers in Zimbabwe.

Teachers' Attitudes Towards Migrants' Children

Teachers view male migrants' children as boastful and extravagant as their parents. They are also seen as rude, wayward, alcoholic, disrespectful, misdirected and not serious with schooling. Teachers treat migrants' children as culprits even where they are victims. They are labelled with all sorts of negatives because of misdeeds of a few.

Teachers connive with children of non-migrant parents to isolate migrants' children as rich, mocking them as children of wealthy people (*abantwana bezikhulu*). Sometimes derogatory labels or names are used on them, for example, children born by cross-border transport operators (*abantwana babomalayitsha*). They are suspected to be children born out of adulterous relationships between *omalayitshas* and their mothers because *omalayitshas* come home more frequently than their fathers.

Girls renting houses at business centres are seen as "food for civil servants" meaning that they prostitute with male civil servants. The civil servants referred to are police officers, soldiers and other government workers in the township. This is because some of the migrants' children who resided there previously had love affairs with some of these civil servants. When girls report cases of thefts and violence by men who force themselves into their rooms, teachers do not sympathise with them but instead mock and insult the girls adding salt to injury.

Attitudes of Migrants' Children Towards Teachers

In revenge learners look down upon teachers, criticising them for being poor, (driving footrons), that is, travelling on foot. They are also mocked for having traditional or dilapidated homes and houses. Teachers' monthly budgets are drawn on the board to mock, embarrass and to prove that teachers have meagre salaries. Some of those who once stayed with parents in South Africa and are from well-to-do families are the most disrespectful of teachers. They do not care about anything or anyone. There are sometimes instances of direct confrontations and fist fights between male migrants' children and teachers.

Migrants' children think teachers are envious of their parents who can afford material goods which are beyond the reach of most teachers in Zimbabwe. Migrant parents afford descent houses, solar or car battery powered lighting system, luxury cars, bicycles, televisions, expensive clothing, quality stationary and groceries for their families. Migrant parents also have livestock, farming implements and can send their children to mission, government and low-cost boarding schools. The *njivas* have a competitive advantage over teachers and other non-migrants working in Zimbabwe.

Learners, especially the males give teachers a tough time by cheating, blackmailing and committing other forms of illicit behaviours. When asked to bring parents or guardians to the disciplinary committee hearing, if caught out of bounds or drinking, they bribe teachers or security guards who caught them in these acts. They also bribe non-relatives to pretend to be guardians. Teachers and security guards who are hard to bribe are blackmailed. Learners investigate into the teachers and guards' private and professional lives and then threaten to expose them if they dare to report their cases *(Tit for tat)*. Such teachers and guards end up being friends to migrants' children defending them before the school authorities and other teachers. Sometimes in a bid to silence learners, male teachers accept the brides.

In cases of shortages, some learners borrow from colleagues while others cheat by borrowing money from friends and teachers lying that they have been instructed by parents to do so. In times of critical shortages, especially in mission and low-cost boarding, learners pretend to be ill or to have forgotten medication so that they are sent home to relatives. They fake diseases as if being demon possessed knowing that for such diseases school authorities would not send them to hospital but home to relatives. They also dodge corporal punishment by pretending to be demon possessed.

Some day-scholars, as a result of lack of educational assistance at home and time for school work, copy homework from others in the morning or pay other learners to do homework for them. Sometimes they share part questions and answers in the morning. Sometimes this is done to fix teachers who give homework to force learners to concentrate on school work. Teachers discover at the end of the term that learners were cheating when they fail dismally despite their good passes in assignments and homework.

Migrants' children boast to teachers that the next time the teachers would meet them the following year they would be driving cars. Those who completed or dropped out of school visit teachers when they return from the diaspora to pose to teachers driving their luxury cars. They switch their radios to full blasts to provoke teachers. Sometimes they ferry teachers to town on weekends just to show off.

The sour relations are now a culture known by everyone in the community. Teachers sometimes face threats of being beaten or stabbed by *njivas*. The relations impact negatively on the respect of teachers by students. Teachers should have good relations with learners for collaboration in education. Enmity tears the two apart and becomes a barrier in schooling of male migrants' children especially. Female learners have better relations with both male and female teachers though there are sometimes incidences where female teachers and grown up repeating female learners clash over *njiva* boyfriends but these are rare.

There role of teachers as *loco-parentis* is jeopardised. The doctrine of loco-parentis is based on the assumption that parents agree to delegate school officials the power of control over the children since the teachers spend the entire day with the children (Nakpodia 2012). Traditionally, this was an arrangement between parents and teachers. The children had no say but today the context of loco-parentis has changed. It has evolved from trilateral model of parents, students and the teacher to include the government which now delegates its powers to children, parents and students (Hall and Mannis 2001). The guaranteeing of children's rights by government and withdrawal of corporal punishment as a form of discipline from both parents and teachers is evidence of government delegation of power. In Bulilima, teachers find it difficult to handle migrants' children amid all the misbehaviour towards teachers. They cannot work with parents because they do not have good relations with parents either.

Traditionally, teachers had a right to search and seize undesirable objects from students' lockers, pockets or even strip search them to ensure safety. Today children have rights to be respected, for example, the right to privacy. Students cannot be disciplined anyhow without parents' consent. Teachers are not often sure where the line should be drawn between the role of parents, teachers and government (Nakpodia 2012). The limited powers of the teacher make resolution of conflicts between teachers and students and issues of discipline problematic. Unresolved conflicts cascade to negative attitudes.

Conclusion

For quality learning to take place, the social environment should be conducive. Social environment has to do with relations between teachers and learners in and outside the classroom, the beliefs and attitudes that learners hold towards teachers vice versa, teachers' relations with the community at large, the totality of relationships (Thamarasseri 2017).The physical, political and economic environment can be excellent but without good and peaceful social relations, learning cannot take place. Cordial parent–teacher partnership is the solution to wars between teachers and learners and between migrant parents and teachers.

References

Baker C et al (2009) The impact of migration on children in the Caribbean. Available from http://www.unicef.org. Accessed 23 Jan 2014

Cortes R (2007) Children and women left behind in labour-sending countries: an appraisal of social risks. UNICEF, New York

Department of Education, Employment and Workplace Relations (2013) Family school partnerships framework: a guide for schools and families. www.familyschool.org/family-school-partnerships-framework.pdf

Filippa OM (2011) Zimbabwe adolescents experience of their parents' absence due to diaspora. UNISA. Available from www.uir.unisa.ac.za. Accessed 12 June 2014

Gavriliuc C (2006) The situation of children left behind by migrating parents. GFMD. Available from http://www.google.co.zw. Accessed 27 Dec 2013

Grazin C, Brigita K (2017) The resolution of conflict between teachers and students: students' narratives. J Teacher Edu Sustain 19(2):107–112

Hall A, Mannis M (2001) In loco-parentis and the professional responsibility of teachers. Waikato J Educ 7:117–128

Kufakurinani U (2014) Transnational parenting and the emergence of diaspora orphans in Zimbabwe, Leiden, Brill: Africa Diaspora

Lasater K (2016) Parent-teacher conflict related to student abilities: the impact on students and family-partnerships. School Community J 26(2):237–262

Nakpodia EG (2012) Teachers responsibility in loco-parentis in secondary schools in Abraka. Metropolis, Delta State, Nigeria. Global J Human Soc Sci 12 (11). Available from http://globaljournals.org. Accessed 22 Aug 2016

Save the Children (2014) Left behind, left out: the impact of migration on children and families of mothers migrating for work abroad. Available from save-sl-leftout. Accessed 12 Dec 2014. https://resourcecentre.savethechildren.net/library/left-behind-left-out-impact-children-and-families-mothers.

Thamarasseri I (2017) Gender, peace and education. A.P.H. Pub Corporation, Cambridge

Tufis A (2007) *The effects of the absence migrant parents on their children school performance.* Soros foundation. Available from http: www.childrenmigration.net. Accessed 12 June 2014

UNICEF (2007) The impact of international migration: children left-behind in selected countries of L. America and the Caribbean. UNICEF. Available from www.unicef.org. Accessed 17 Jan 2014

Wei Lu G (2011) Left behind children in rural China: research based on the use of qualitative methods in inner Mongolia. University of New York, New York

Worley V (2003) The teachers place in the moral equation: in loco-parentis. Oklahoma State University, Still Water

Chapter 10
Intervention Strategies for Equity in Education: Synchronised and Co-ordinated Model to Help Learners Cope with Absence of Parents Due to Migration in Zimbabwe

Western Intervention Strategies for Equity in Education

The dominant western ways of thinking in education are that "one size fits all" (Florian 2012). Western intervention policies in education ignore marginalisation, diversity, educational settings and contextual differences yet education is supposed to be a panacea of the world's problems in different settings (Watton 2018). Western policies ignore the history of colonisation, socio-economic inequality and environmental precarity yet the traces of colonisation still remain and continue to shape conditions of member countries like Zimbabwe.

Despite the introduction of equity policies through western intervention strategies in education, learners are still deprived because these equity policies do not go further to cater for various background circumstances associated with poverty, discrimination, marginalisation and vulnerability. There has been declaration on equity, access, participation and social justice for all learners in different educational contexts but a large number of learners still do not have access to basic education. They drop out of school or are discriminated within the schools (Mukhopadhyay 2015).

In spite of declarations of equity policies, dropout rates are still high in Zimbabwe and Africa, retention of children in schooling is still a major concern. Inoue et al. (2015) reports that about 89 million youths are out of school worldwide which is half of youths in Sub-Saharan Africa. Around 42% of African school-going children drop out before end of primary school. It is estimated that 40million youths will drop out in Africa and this is a potential explosive economic and social disaster. The countries in Africa with low retention rates are Burundi, the two Congos, Lesotho, Zimbabwe, Tanzania, Madagascar, Niger, Rwanda and Sao Tome. In Southern Africa, Malawi has the highest dropout rate of 15% girls and 12% boys. These are low-income countries plagued with poverty, early marriages, migration, low parental education, lack of access to education and some are war-torn countries.

Kageler (2015) notes that in poorest countries and households, learners are likely to be out of school especially where there are no social safety nets to cover for chil-

W. Matsa, *Marginality, Migration and Education*,
https://doi.org/10.1007/978-3-030-60873-6_10

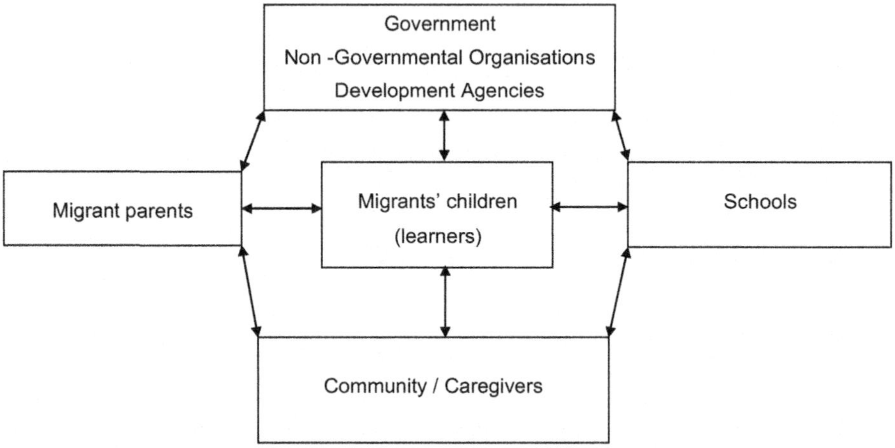

Fig 10.1 Synchronised and co-ordinated coping strategy model

dren of different economic backgrounds. In Zimbabwe, dropout rates are higher at secondary than primary school. They are higher for girls than boys. About 30% of girls drop out in forms 3 and 4. Reasons for dropping out are largely failure to pay fees though these vary with provinces, ages and gender (Kageler 2015). Since 2012, enrolments have gone down, the numbers for out of school children have risen to 1,234,641 and most of these are orphans and the disabled though the category of marginalised and vulnerable children is broader. It includes even migrants' children and those from poverty-stricken backgrounds.

Financial Gazette of March 12 (2015) reports that dropping out of school in 2013 rose to 13,000, which is 40% from 2012. Boys contributed the biggest number 7083 and girls 6170. Bulawayo 24 News of October 2015 also echos that a total of 1572 secondary school pupils in Matabeleland South province dropped during term one. Matobo district recorded 654, Beitbridge 634, Insiza 509, Umzingwane 92, Mangwe 582 and Bulilima 57. Bulilima and Mangwe are not spared of this plague. Reasons for dropping out are province or district specific. In Matabeleland South province, Bulilima and Mangwe districts dropout result from poverty, hunger, lack of parental guidance due to migration, early marriages to njivas and desire to trek to Botswana and South Africa deemed lands of "milk and honey". Children envy those already working there having flashy cars and enticing fashion (Moyo et al. 2016, p. 860). Bulawyo 24 News (10 October 2014) has it that 10 girls dropped out of Tokwana high school because of pregnancy. About 71 stopped going to school after getting married to njivas. They fail to continue after pregnancy because of stigma.

Poverty is another major factor in learning and retention in school. Lack of fees being the main cause for dropping out. School attendance relate to poverty, level of education for the household and early marriages (Kageler 2015). From 2012, BEAM funding was reduced and this kept the vulnerable out of school. Girls were assisted by CAMFED, World Vision and higher life but the numbered sored because of cultural beliefs of son preference when it comes to fee payment for schooling by parents and guardians.

Equity in education means that personal or social circumstances such as gender, ethnicity, family and background are not obstacles to achieving educational potential, fairness and that all individuals reach at least a minimum level of skills despite the socio-economic circumstances (OECD 2012). A number of equity policies were introduced in the education system in Zimbabwe some of which are outright western interventions stemming from international declarations.

Educational needs in Africa still remain the greatest despite Education for All (EFA) which was a global and local equity policy. Global monitoring report has it that the net enrolment of students during the EFA rose significantly to 20% from 1992 to 2012 in 17 countries, 11 of which are in Sub-Saharan Africa. This was remarkably in the region with political and economic challenges. However, the increase in enrolments in public schools led to decrease in qualifications and teacher–pupil ratios (Stephen 2015). A solution to one problem gave rise to another. EFA meant that everyone had a right education despite race, gender, ethnicity, religion, disability and locality among others. This was in line with the Universal declaration of Human Rights (HR), a global declaration.

EFA has origins in the Jomtien Conference in Thailand 1990, a western intervention strategy in education. It was re-affirmed at the 2000 World Education Forum in Dakar, Senegal. In Zimbabwe, it was introduced at independence accompanied by free and compulsory universal education. To cope with the rapid growth in student enrolments, states hired young and unqualified teachers without training or employment contracts. This resulted in a drop in quality of education in public schools and the rise of private schools with sufficient facilities and resources. Private schools were accompanied by private tutoring. The EFA equity policy assumed that schools had sufficient infrastructure, capital and human resources. Learners' problems now shifted from the exterior to the interior of the school. Learners' problems were no longer those of lack of access but problems related to factors of socio-economic status of parents, marginality of regions, discrimination and generally unfair ground. Children of the poor continued to be discriminated in private schools with better resources. They accessed poor and marginalised schools lacking modern facilities, technology and qualified teachers with specialised training. Stephen (2015) noted that concerns about equity in education do not start and end in the classroom. It starts with one's background at home and has ripple effects to employment and adult life. Educational background, opportunity and achievement established at childhood persist into adulthood resulting from social inequalities and cumulative disadvantages. Education for All was accompanied by free primary education a universal policy at least in the elementary and fundamental stage. Free primary education meant that children accessed education without or with minimal payment for levies. However, other forms of inequality, poverty, class and gender remained a barrier to equity in education.

Universal primary education represents universalistic view of what primary education should be like. It is based on western conceptions of education. Its justification is based on statistical evidence that one quarter of children not attending school live in the Sub-Saharan Africa, South West Asia, South America and Caribbean, the poor to justify the imposition of such policies on the South. The policies are laden with their values and epistemologies. The policies do not transform the education systems of the former colonies nor do they give epistemic freedoms. It is a way of denying non-western countries education growing out of their environments (Tickly 2001). Therefore, post-colonial theory of education deconstructs the context of global educational partnerships and dismisses it as a form of imperialism and marginalisation of the non-western countries. Western intervention strategies perpetuate historical and economic disparities.

In the 1990s, inclusive education with equitable educational opportunity was introduced in Zimbabwe although elsewhere, it dates back to the 1980s. Inclusive education is viewed as a western concept associated with neo-colonialism with its own definitions of the excluded and marginalised (Mukhopadhyay 2015). Inclusive education seeks to reduce exclusion from and within schools, secure participation and learning success for all although it is criticised as neo-colonialist concept and an unwelcome imposition (Watton 2018). It seeks social justice by resisting exclusion from within school communities, promoting access, participation and achievement of all learners. It was also meant to reduce segregation and overcome barriers to education. Historically, charities and religionists provided services and not national governments. It was based on the model that categorised the excluded as those with disabilities. Up to today, the model follows an integrationist where those with disabilities are mainstreamed into existing settings. The settings of the school do not change; hence, it is more adaptive than transformatory.

However, by confining it to learners with special needs or disabilities, other children experiencing other forms of marginality and vulnerability are excluded. Inclusive education depends and continues to rely on donor funding from international agencies. Generally, western intervention in equity in education rely on international funders, UNICEF, UNESCO, Department of international development (DFID), Swedish Development Cooperation Agency (SIDA), WB and DANIDA among others. Other donors are non-governmental organisations also with western concepts of marginality, vulnerability and western concept of quality education. Western intervention strategies by donors do not cover a wholesome child. For example, UNESCO focuses on educators, UNICEF-educational materials, Plan international-educational infrastructure, Higher Life-fees ignoring other factors like poverty at home, living and schooling environments, health, educational background of parents and caregivers, political, social and economic marginalisation. To make matters worse, donor organisation operate in different regions focusing on their own specific factors of access and quality of education. African governments, Zimbabwe included, rely on funding.

External aid contributes a significant share most of it being donor funds. About 46% of public funds in low-income countries is from external aid (UNICEF 2015). African countries are baffled by fragile political, economic and social challenges. These intersect with the regions' vulnerability to environmental disasters and post-colonial inequalities as well as lack of social safety nets. Dependency on donor aid has a negative impact on achievements of equity policies in education.

Millennium Development Goals (Goal, 2000–2015) and Sustainable Development Goals (2015–2030) are all meant to back up equity policies in education. However, these have had their share of problems relating to discrimination and marginalisation due to governances, health of economies and budgetary allocations. Universalising the understanding and implementation of policies in diverse contexts laden with various forms of inequality and vulnerability is, in fact, perpetuating the existing discrimination and marginalisation.

Western intervention policies in education reflect inequalities between North and South, perpetuates issues of poverty and control and conditions of oppression in an invisible way (Martin 2012). For example, education is funded, the success of the clause in the MDGs and SDGs on education, depends on funding from outside. The problems identified in relation to each MDG relate more to the South than the North. The west through funding is then viewed as advancing society in a paternalistic way. The MDGs as an intervention strategy become a gift from the North. Those on the South do not play an active role in the achievement of the goals (Martin 2012). Almost all financial problems related to education in the south are funded through AID by western governments through global international, non-governmental organisations like UNICEF, UNESCO, PLAN International and Oxfam among others. The aim is to influence pupils' learning, their content, pedagogy and the general purpose of education, to maintain economic, political and cultural imperialism through education systems that perpetuate these inequalities. The inequalities are hidden under the concept of care, help, aid, maintaining helplessness of others (Andreotti 2008, p. 59)

The recently Early Childhood Education/Early Childhood Development policy started in United States and spread to all public policy institutions long before 2000 to help curb delays in individual learning capacity. It is derived from European and North American traditions based on Euro-centric perspective of children's rights. The policy also universalises children (ILO 2012).

The gender clauses in the Zimbabwe National Gender Policy (ZNGP), Gender Affirmative Action Policy (GAAP), Sexual Harassment Policy and Pregnancy Policy in Zimbabwe are all gender equity policies meant to mainstream gender in education. However, they were imposed on existing structures of gender and colonial inequalities. Existing social, cultural and economic arrangements of society have not been changed. They still serve the interests of patriarchy and the dominant classes in society making it difficult to achieve equity in education. All the educational policies meant to bring equity in Zimbabwe have a western element of ignoring the existing power structures of inequality. Integrating changes within existing forces of discrimination and power inequalities does not bring equity.

Post-colonial Theory and Education

Post-colonial theory and its expositions have its roots in colonial education system which sponsored assimilation of European content and valorised local realities of non-western societies. Colonial education was seen by Walter Rodney as an organised form of imperialism (Ndile 2018). Post-colonial theory questions the post-colonial education system as incarnation or continuation of old habits, structure and attitudes of colonialism to further cultural oppression (Hapanyengwi-Chemhuru and Makuvaza 2017). Post-colonial education theory probes the impact of cultural, economic, political, psychological and military dominance perpetuated by education of the West in Africa and other former colonies. Post-colonial concept of education draws heavily from post-structuralists' notions of power and knowledge discourse. Formal education and its globalised nature is a form of marginalisation. It allows for the exertion of power to spread western hegemony through control of knowledge. Formal education represents inherent European values at the expense of African ones. It results in irrelevant education for non-western countries (Tickly 2001). This is done in an invisible way through formal education.

According to post-colonial theory, education in post-colonial societies, the curriculum and content is determined by the aims of education which do not differ from those of the colonial period. The purpose of education during colonisation was to eradicate African cultures, histories and knowledge. This kind of education is brought by knowers and does not initiate individuals into the general culture of their communities, but the experiences of the west. Learning is not related to the pattern of life of Africans but privileges western cultures, experiences and epistemologies (Ndile 2018).

Post-colonial theory attempts to re-asset epistemological values of non-European world (Ghandi 1998:40). It is an attempt to retrieve the so-called illegitimate or disqualified and subjugated knowledge of the former colonies. It questions hegemony of European dominance of western styles of knowledge which legitimise and sustains asymmetrical power relations associated with processes of exclusion and dominance (Osmar 2012).

Post-colonial theory calls for an epistemological decoloniality (Quijano 2007), which is deconstruction of western concepts of knowledge, power, culture, politics and education policies that drive the curriculum content and the pedagogy in education. Post-colonial theory challenges notions of globalisation, international partnerships in education and universal education as neutral. All these are new forms of imperialism meant to control the direction of development and education in non-western countries (Lavia 2007). Global partnership policies in education are hegemonic discourses. Western knowledge and academy is at the centre of policy and practice. Partnerships help children of the South (colonised) to learn experiences of children of the North (colonisers), their cultures and traditions (DFID 2007, pp. 5–7). Global school partnerships reinforce colonial patterns of behaviour, the very stereotypes and attitudes they pretend to change (Martin 2012).

Post-colonial Intervention Strategies in Education

Post-colonial strategies of intervention in education are a reaction to western intervention strategies that seem to perpetuate neo-colonialism and its inequalities. Post-colonial theory looks at the relationship between the colonised and colonisers, the position of dominant and the dominated, the perspectives of the dominated, marginalised, oppressed and subordinated. (Westhuizen 2014). In this case, the former colonised and former colonisers with their dominant education philosophies and equity policies. It is about giving privileges to the voices of the marginalised, identifying and correcting issues affecting the marginalised. It is a fight against mainstream methods that reinforce stereotypes.

This can be done through social transformation through policy research for social change. Policy research and influence is a central process in making migrants and other vulnerable children visible. They need to be conceptualised as vulnerable, marginalised, discriminated and steps are to be taken to include them in policy planning and implementation if inclusive education, equity and quality education is to be achieved.

Complexities inherent in post-colonial conditions need to be analysed. Migrants' children in Zimbabwe, in particular, Bulilima and Mangwe, the hot spot of migration to Botswana and SA are geo-physically, politically, ethnically, economically and culturally marginalised. They experience various forms of marginality. Teachers and caregivers lack skills of handling migrants' children because they lack empowerment. In implementing equity policies to achieve quality education, the socio-economic and cultural context should be kept in mind.

Beside legal and policy regimes streaming from the international declarations, EFA, ECD, universal primary education, inclusive education among others, at local level the Zimbabwean government added some to create an enabling environment for quality education. Under school health policy of 2014, life skills, sexuality, HIV and AIDS education strategic plan was launched between 2012 and 2015 in line with the MDGS. This was meant to retain adolescents at school through prevention of infection and positive living to pave way for acceptance of Anti-Retro-Viral drug use even in schools. Pregnancy policy was meant to improve retention of girls to increase completion of secondary education. Non-formal education policy complements other policies to ensure that out-of-school children complete their secondary education.

Education Medium Term plan stretched from 2011 to 2015. It focused on the teacher–pupils ratios, expected to reduce them to 1:28, pupil–textbook ratio of 1:1 and 30% budget allocation to education, gender parity in enrolment as well as increasing literacy rates from 88.4% to 98% by 2015. This was to improve the quality of learning, learning conditions and resources. In 2014, UNICEF distributed text and exercise books to reduce the ratios. The Amendment of the constitution no. 20, 2013 provides for section 19 on children's rights and section 22 on education. However, poverty levels compromise the milestones made by the policies and legal instruments.

Vulnerable and marginalised children suffer high levels of exclusion even in implementation of equity programmes for quality education at each stage of enrolment, completion converges to marginalise, discriminate and deprive them. As a result, the coping strategies by migrants' children are adaptive rather than transformatory, reactive rather than proactive. Children's coping strategies consisted of adaptation or survival strategies. There are no guiding policies. Above all, the strategies do not go deeper to eliminate the challenges but only minimise their effects. The coping strategies are fragmented with each group dealing with its own challenges. The focus is not on the vulnerable group of migrants' children as a category.

Existing Fragmented and Adaptive Coping Strategies

Each group of teachers, migrant parents, caregivers and learners focused on challenges that affected them as an individual group. Issues of diversity are ignored. Teachers and caregivers do not seem to care about challenges experienced by migrants' children but only those that affect them directly. Migrant parents are not aware of the extent of the effects of the challenges on their children or have no solutions.

The coping strategies used in Zimbabwe are divided into two; those that operate at four institutional levels, the government, non-governmental organisations and development agencies. The second level is that of schools, third the community level and the fourth the individual at family level but all in a context laden with inequalities and discrimination. However, the four levels need to work in partnership.

In Zimbabwe, in Bulilima and Mangwe, there is no deliberate effort by government, school authorities or the community to ensure that CRCs are enforced or respected. There are no organisations advocating for children left behind by migrant. Government policies and legal framework for children left behind is missing. Migrant parents and caregivers are not bound by any restrictions of policy or law. Teachers, caregivers, migrant parents and migrant children's coping strategies at home and school in Bulilima and Mangwe are reactive rather than pro-active. Moreover, the reactive strategies are passive rather than transformatory. The coping mechanism by learners, teachers and caregivers do not alleviate challenges but only minimise the negative effects of migration in schooling.

Of interest to note is that learners think their challenges have no solutions. Teachers, migrant parents and caregivers do not seem to have solutions for children's challenges either. At institutional levels, nothing of government involvement is in place to alleviate the plight of migrants' children. No laws have been lobbied by the government on behalf of disadvantaged children. Marginalisation starts at national level. In an attempt to assist migrants' children to cope with educational and emotional challenges, counselling sessions are carried out by class teachers, invited chaplains, superintendents and school disciplinary committees. However, the counselling sessions do not target challenges of migrants' children but challenges that affect all learners. Migrants' children are not conceptualised as a special group with challenges specific to them. Invited guest speakers and teachers have guidance and counselling lessons that do not target the realities of children with absent parents due to migration.

Social welfare programmes by the government and NGOs borrow from western intervention strategies and do not target migrants' children as vulnerable children. They are classified and ignored as rich, down playing other forms of diversity that make them vulnerable. Moreover, not all migrant parents remitted sufficiently. There is need for a shift by government and NGOs, for example, Plan International, Capernaum (Higher Life), Social Welfare and Basic Education Assistant Module (BEAM) to targeting migrants' children as a separate category at risk and deserving benefits from the above schemes. The existing coping strategies are for inclusion of migrants' children into already existing general programmes. No programmes are specifically set up for them.

Expansion of mission, government and low-cost boarding facilities and establishment of more in those schools or areas that do not have any is seen as one of the solutions to the plight of migrants' children failing to access schooling. Low-cost boarding is affordable and could benefit more migrants' children in Bulilima and Mangwe districts. Boarding facilities not only ensure protection of children with absent parents from harm, but also ensure access to quality education. However, accessing these depends on financial and material capital. As long as they are not part of categories benefitting from government and non-governmental grants, scholarships, welfare programmes, some from poor families and non-remitting parents will remain excluded. The affordable existing low-cost facilities are of poor quality compared to mission boarding schools. They, therefore, do not close the gap for absent parents and lack of educational and moral support from home.

Setting up school-based or mobile chaplain's office for counselling of migrants' children during guidance and counselling lessons as in established boarding schools can be of help. Instead of complaints directed to teachers and school authorities who are sometimes culprits themselves, the chaplain would be expected to attend to these. Chaplains, like social workers, are crucial in attending to children's psychological needs and in imparting skills of handling psychological problems. However, this can only be effective if approached from perspectives of vulnerability of migrants' children, understanding their plight as a separate category with challenges specific to them.

Coping strategies that operate within the existing systems of oppression, exclusion and inequality reproduce the same cycle of oppression, exclusion and marginalisation. Some of the strategies adopted by learners to adapt to educational challenges do not go far enough to solve challenges but only postpone the challenges. For example, borrowing money, food and educational resources, theft, violence, paying others to do homework, keeping quiet and ignoring everything not only affected the quality of education but also produced future behavioural problems. Teachers and caregivers do not assist because they have negative attitudes towards migrants' children.

The "lacks" experienced by migrants' children and their dire need of human support suggest lack of empowerment of learners themselves, teachers, caregivers and migrant parents as well as the community. Empowerment of all stakeholders can create good relations and curb the blame game between teachers and learners, migrant parents and teachers and finally between migrants' children and caregivers.

Empowerment needs are to be broadened to include the policy and legal framework on child protection. Schools can sensitise the community parents and children on the importance of education and educational challenges of migrants' children on annual general meetings, consultation days and prize-giving days. So far, there are no deliberate efforts to sensitise or counsel teachers and migrant parents. Counselling programmes by school authorities, class teachers, superintendents, pastors, police officers and invited guests are all directed at learners ignoring teachers, migrant parents or caregivers who are sometimes the culprits of ill-treatment or lack of empowerment.

Educational awareness campaigns, workshops and seminars can be carried out on improved communication technology which is cheaper, simple and adaptable to rural areas. Educational campaigns encourage greater contact through cell phones, Skype and emails. However, emails and Skype are not adaptable to poor and marginalised rural areas.

Coping Strategies at Home

Most of the strategies suggested hinge on the need for human support; for example, reporting to migrant parents, relatives, neighbours, police officers and asking for educational assistance from teachers, friends, brothers and sisters. Otherwise, most learners bottled up their problems or find ways of adapting to the situation as a survival strategy.

Of interest to note is that learners feel that their parents or caregivers are the ones to solve problems for them while others think their challenges have no solutions. They just have to endure. Migrants' children are the ones to change to fit into the situation. Migrant parents' hands are tied by absence and distance. They are quoted between sacrificing their jobs and their children's education.

Coping Strategies at School

The coping strategies by learners at school are not different from those that learners apply at home. The strategies only help learners to survive in the school environment. Nothing is done at school to recognise them as a peculiar group with peculiar problems. They are treated as if all things are equal. Nothing is done in relation to ill-treatment by teachers; it is taken to be the norm. Their parents have no power over teachers. They desire to be put their children in mission, government or low-cost boarding, and for those who cannot, at least to foster them with trusted relatives, who have the best interest of the child at heart.

Coping Strategies by Teachers

There are guidance and counselling lessons every week or anytime when the need arise though these are also not specific to problems of migrants' children related to migration. Class teachers also counsel students during class teachers' registration period. Children's close relatives are invited or parents are phoned by teachers to discuss the children's challenges and to counsel both children and parents. Teachers are not the best to counsel because some have negative attitudes towards migrant parents and their children. However, this happens when serious cases arise. This cannot be a norm because it is expensive. School authorities are sometimes culprits in these challenges. It is a situation where a problem causer tries to be a problem solver.

Local pastors of particular denominations are invited in mission, government and low-cost boarding schools to talk to learners about especially academic and social problems they encounter at school. Boarding superintendents and matrons also have times set aside to counsel learners. However, their talks are not specific to migrants' children. Matrons, superintendents, school authorities and teachers act *in loco-parentis*, handling issues which are to be handled by parents and caregivers. They act as link persons between migrant parents and caregivers.

Invitations of chaplains, peer educators, the police to discuss challenges encountered by learners with absent parents is seen as handy though the police tend to focus on crime, violence and abuse in general. There is need to coordinate with police sensitising units to prepare programmes that incorporate problems of children with absent parents. Strategies need to be transformed to respond to specific needs of children with absent parents.

There are no well-established government boarding schools but only mission boarding schools which are expensive courtesy of Methodist, Seventh Day Adventist and UCCSA (traditionally known as London missionary society), Roman Catholic, Anglican and Salvation Army mission schools, some of which are single sex schools. Mission schools accept students on payment plans and give first preference to learners from their denominations. However, low-cost government boarding schools disregard gender, faith and class giving all learners equal opportunities to education.

Communities and schools in Bulilima have set up child protection committees to focus on academic challenges and the overall welfare of migrants' children because in these districts. Almost every family cares for children of migrants at primary or secondary school. The problem is that migrants' children are not seen as having problems specific to them as children with absent parents. The perspective of teachers, community, non-governmental organisations (NGOs) and the government should shift from viewing children of migrants as rich, when, in fact, they are vulnerable children. The social welfare and Basic Education Assistance Module (BEAM) by government and other programmes of NGOs, for example, Plan International, World Vision and (Higher Life) Capernaum focus on orphans, double orphans, HIV-affected and infected, physically and mentally challenged, those with ill parents or imprisoned parents, but leave out children with absent parents due to migration. Not all migrants' children are well catered for. The selection criteria by community schools' selection committees do not target migrants' children.

Synchronised and Co-ordinated Model to Help Learners Cope with the Impact of Absent Parents Due to Migration

Recommendations should start at national level cascading down to the family level.

Government, NGOs and Development Agencies Level

The government should come up with a policy and legal framework on migration of parents who live children behind. These policies and laws should take into consideration CRCs and laws on the rights of children. The policies and legal framework should bind and direct migrating parents, teachers and caregivers.

There should also be deliberate policies by the government and NGOs to target migrants' children as a vulnerable group as happens to orphans and those infected and affected by HIV and AIDS among others. Migrants' children have their unique experiences and challenges that need attention. The BEAM (Basic Education Assistance Module) by government caters for the poor and the children of ex-combatants.

Regional or District Directors in Education and education officers in Bulilima and Mangwe need to launch educational awareness campaigns to conscientise Bulilima and Mangwe communities on the importance of education, the dangers of parental migration on schooling, child development and the violation of children's rights. The Ministry of Education should work in partnership with NGOs and school Development Committees to expand boarding facilities in Bulilima and Mangwe. Some schools do not have conventional boarding or formalised low-cost facilities while others offer less formalised low-cost boarding that are not secure. The government is expected to embrace these and supervise them to ensure respect of the rights of children.

Migration has a negative impact on schooling of migrants' children in not only Bulilima and Mangwe, but also the whole region of Matabeleland South and probably the whole of Zimbabwe. Regional directors in Education and Education officers working in partnership with schools and their school development associations should hold training workshops and seminars to deconstruct the belief that all migrant parents and their children are wealthy people with no challenges. Communities should be made aware of the fact that migrants' children are vulnerable children and carers need skills in handling them.

The Ministry of Education projects and programmes in schools need to take experiences and challenges of migrants' children into account. In schools, there are education programmes on sex education, reproductive rights, reproductive health, and HIV and AIDS education. Problems of migrants' children are not part of these programmes.

School Level

The school authorities in partnership with School Development Associations need to launch community sensitisation programmes on the importance of education, parental guidance and the challenges experienced by migrants' children. The programmes can be successfully promoted during prize-giving days, consultation days, Annual General Meetings and any other important days on the school calendar. This can help make the community aware of the challenges and vulnerability of these children to social ills.

School should also establish child protection committees that work with school authorities and community child protection committees. Some of the schools have these, but they are limited to the school horizon yet migrants' children have challenges both at home and school. Most of the school challenges emanate from home. School child protection committees should then work in liaison with the community, migrant parents and caregivers.

It is crucial for schools to hold meetings with migrant parents and caregivers to discuss the plight of left-behind children to enlighten especially migrant parents on the challenges experienced by their left-behind children at home and school. Otherwise, they depend on unreliable and segmented information from their children and caregivers. There is need for a holistic approach. School authorities should also forge for a strong partnership between migrant parents, caregivers and teachers for cordial relations and effective care for children left behind.

At Provincial and District Education offices in Zimbabwe, there are education officers available for psychological and counselling services to cater for teachers and learners. Individual schools can lobby for the appointment of school-based social workers or school-based chaplains as that happens in private schools. They can work with the provincial and district psychological and counselling services for the welfare of migrants' children. They can also handle all the complaints and grievances of left-behind children, instead of teachers who are overwhelmed by a lot of work. For those schools that cannot pay a social worker or school-based chaplain, they can form clusters so that the social workers or the chaplain's offices can attend to children's welfare on specific days.

In schools, there are peer empowerment programmes, but these tend to focus on drug and alcohol abuse, HIV and AIDS, reproductive health, being orphaned and other vulnerabilities. Although migrants' children do benefit from these programmes, some of their challenges specific to them are left out. These programmes need to be broadened to include challenges of migrants' children.

School authorities should have dialogue with migrant parents and encourage frequent visits and communication during stipulated times and dates instead of discouraging it. Banning communication between the two enlarges the gap of detachment.

For delayed fees and other educational resources, schools and migrant parents should draw up or negotiate payment plans that give migrant parents and their children a one-month allowance before children are sent home for non-payment of fees. In Zimbabwe, learners whose fees are paid by government as government grants, BEAM social welfare or non-governmental organisation grants are not sent home for fees but schools wait until the government or NGOs pay. Even tertiary institutions, colleges and universities have payment plans drawn and agreed between the institution and the parent or guardian.

The NGOs offering welfare services in the area, for example, World Vision should shift from an "add-on strategy" to isolate and target migrants' children as a unique group with educational and other challenges different from those of other vulnerable groups of children. Some of these organisations assist in basic food, others in fees, medical care and scholarships. The focus to date has been on single orphans, double orphans and children infected or affected by HIV and AIDS.

Community Level

The community through the school Development and Teacher Associations should try to advocate for inclusion of migrants' children with serious challenges in welfare assistance schemes by government and NGOs. School Development Committees are part of selection committees for other vulnerable children. They should also select children who are vulnerable due to migration of parents. Some of these children are "orphaned" infected and affected by migration of parents.

It is important for school Development Committees to mobilise resources from their communities, seek donations from government and other well-wishers to expand formalised low-cost boarding, facilities to cater for more migrants' children. Non-formalised low-cost boarding needs to be formalised and expanded also, for the security of children.

Communities should also establish child protection committees who can work with security personnel and school child protection committees. These should partner with police officers and neighbourhood watch personnel to monitor the welfare of migrants' children at home and school. The security should be compliant with the law and the rights of children to address issues of insecurity and lack of protection of child renters and other children in general.

Family Level

After having been sensitised and conscientised by various stakeholders on the challenges of migrants' children, families should be able to create a conducive environment for schooling of migrants' children.

Migrant parents need to discuss intentions, negotiate and explain the importance of migration to family members including their children.

Care or foster arrangements must be chosen by migrant parents before migrating taking into account the best interests of the child. Migrant parents should visit and phone more frequently to create a strong bond between them and their children.

Migrant parents should take it upon themselves to provide basic needs and educational needs not only for their children but also for carers to reduce economic strain on caregivers.

Migrating parents, in particular, need empowerment on CRC, for example, Article 9 discouraging separation of children from parents and the best interest of the child, Article 28, placing value on education, Article 5 on importance of parental guidance and Article 3 on protection from harm.

The coping strategies in Bulilima and Mangwe are fragmented. Each institution and individual groups work separately. Schools, families and the community have diverse strategies that are not co-ordinated. The strategies lack a formidable united force for the benefit of vulnerable migrants' children. Below is a suggested synchronised and co-ordinated model for coping strategies by different stakeholders in Bulilima and Mangwe districts.

No group should work in isolation and migrants' children should be at the centre and benefit to a greater extent. All the stakeholders should work in partnership for the welfare of migrants' children. Working in isolation does not achieve the desired results. Government and non-governmental strategies should filter down through all institutions to the grass-roots level for the benefit of learners who are vulnerable children of migrants. The national level coping strategies should be interlinked with coping strategies at other institutional levels like non-governmental organisations, schools, community and the family level.

Conclusion

However, the model can be more effective if all the strategies are transformatory and are migrants' children specific. Shift in conceptualisation of these children and the development of a policy framework to help solve their plight is the starting point. All stakeholders should join hands and work in partnership. Policy and legal regimes should be instituted to track and monitor migration, non-completion rates and causal factors and to enforce policies and laws that protect migrants' children. Existing acts and policies are to identify the current body of legislation to ensure it agrees with the constitution. The constitution hammers on respect of children's rights, hence educational policies should monitor observance of the rights of children in education. The legislation that does not conform to the demands of the constitution needs to be amended. For example, section 22 of the constitution provides for free and compulsory education for children but education in Zimbabwe is not free. Legal and policy provisions need to be transformed into action for children with focus on poverty, child marriages, health and nutrition, violation of rights, educational funding and legal and policy monitoring (Kageler 2015). Tangible low-cost interventions are also very crucial, for example, feeding schemes, community bursaries, scholarships, health education, sexuality and reproductive rights education, career guidance and guidance and counselling to enable vulnerable migrants' children to cope with challenges of absent parents due to migration.

References

Andreotti V (2008) Development vs poverty: Notions of cultural supremacy in development education policy. In: Bourn D (ed) Development education: debates and dialogues. Institute of Education, London

DFID (2007) The world classroom: developing global partnerships in education. DFID, London

Florian LC (2012) Preparing teachers to work in inclusive classrooms: key lessons for the professional development of teacher education from Scotland inclusive practice project. J Teach Educ 63(4):275–285

Ghandi L (1998) Postcolonial theory: a critical introduction. Colombia University Press, New York

Hapanyengwi-Chemhuru O, Makuvaza N (2017) Re-thinking education in postcolonial in Sub-Saharan Africa in the 21st century. Sense Publishers, Rotterdam, pp 85–103

ILO (2012). Right beginnings: ECE and Educators. Global Forum Discussions 22–23 Feb 2012 report. ILO, Geneva

Inoue K et al (2015) Out-of-school youth in Sub-Saharan Africa: a policy perspective. Worldbank, Washington

Kageler S (2015) Education sector analysis Zimbabwe: final report. https://www.globalpartnership.

Lavia J (2007) Repositioning pedagogies and postcolonialism: theories, contradictions and possibilities. Int J Incl Educ *11*(3):283–300

Martin FE (2012) Power and representation: a postcolonial reading of global partnerships and teacher development through North-South studies. UKAID. https://core.ac.uk. Accessed 28 Apr 2020.

Moyo S, Ncube D, Khupe M (2016) An assessment of factors contributing to high secondary school pupils' dropout rates in Zimbabwe: a case of Bulilima district. Global J Adv Res 3(9):855–863

Mukhopadhyay S (2015) West is best? A post-colonial perspective on the implementation of inclusive education in Botswana. Kedi J Educ Policy 2015:19–39

Ndile RR (2018) Education transformation in post-colonial independent Africa: a historical assessment of the Africanisation project. Centre for Educational Transformation, Auckland

OECD (2012) Equity and equality in education: support the disadvantaged students in schools, OECD Publications. Available from www.oecd.org/czch/49603567.pdf. Accessed 28 Feb 2020

Osmar SM (2012) Re-thinking development from a post-colonial perspective. J Conflictol 3(1):42–48

Quijano A (2007) Coloniality, modernity and rationality. Cult Stud 2(2):168–178

Stephen R (2015) The long and short term views on equity in education: steady progress or whack-a-mole? Int Rev Educ 61:439–444

The Financial Gazette (2015) Over 13 000 pupils drop out of school. www.financialgazette.cozw.

Tickly L (2001) Globalisation and education in the post-colonial world: towards a conceptual framework. Comp educ 37(2):151–171

UNICEF (2015) Education for all 2000–2015: achievements and challenges. EFA global monitoring report 2015. UNESCO, Paris

Watton E (2018) Educational research for social change. Acad J 7:31–45

Westhuizen GJ (2014) Post-colonial perspectives on education policy research. Afr Educ Rev 10(4):689–708

Chapter 11
Implications of the Book

The book shows interplay of marginality, migration and education. It has implications to policy makers on migration and education, migration and development, gender and migration, migration and children's rights and marginality and education among others. The book suggests a look into issues of geo-physical, socio-economic, political and cultural marginalisation. It directs focus on a variety of forms of inequality and unequal access and provisions of resources resulting from state of marginality of Bulilima and Mangwe people of Zimbabwe with particular focus on migrants' children. Although the book focuses mainly on marginalisation of migrants' children, the Zimbabwe community as a whole experiences marginality. Non-migrant children live in marginal contexts also although their situation cannot be compared to that of migrants' children.

Poverty and gender are intervening variables to worsen marginality of the families compromising child development and schooling of migrants' children. The impact of marginality and migration on left-behind children depend on the gender of the migrant, caregiver and the gender of left-behind child. Implications to challenges in schooling on migrants' children depend on the gendered experiences of the child. The book then suggests mainstreaming of gender and inclusiveness in migration, education, and development. It redirects governments, NGOs and development agencies as well as local communities to focus on risky schooling environments right from physical, socio-economic, political and cultural ones to remove barriers to learning to make them gender responsive. The book problematises the relationship between marginality, migration and education emphasising educational experiences of migrants' children in the face of marginality and migration in Zimbabwe.

The book suggests a twist to reconceptualisation of marginalised vulnerable children affected by various forms of inequality and exclusion. It lobbies for reconceptualisation of migrants' children as marginalised and vulnerable category to be considered so in policy making. The definition of the vulnerable need not be limited to orphans and the disabled. It should be broadened to include other groups, even

W. Matsa, *Marginality, Migration and Education*,
https://doi.org/10.1007/978-3-030-60873-6_11

migrants' children. It also challenges traditional views of seeing children as passive victims rather than active participants in migratory processes. There should be rethinking of children as independent moral subjects and not appendages of parents when it comes to migratory debates and policies as suggested by Jesefsson (2014).

There has been lack of focus on left-behind category of children, adolescents with their unique problems as well as focus on children as independent from migrating parents in both policy and practice. Migrants' children are "policy orphans" positioned in no mans' land with little data to develop evidence-based advocacy framework (Inoue et al. 2015, p. 1). There is no coordination of ministries of labour, education, health and child welfare in dealing with migrants' children. There is need of policy mix to link labour, migration policies and well-being of children to ensure that children left behind can reach their potential. The general impacts have also not been seen in the context of poverty, vulnerability, gender and children's rights (UNICEF 2007). The book considers this intricate relationship. The adolescents should matter in migratory debates because they are a vulnerable group even before migration of parents. They should be treated as special with challenges to be prioritised.

The book also implies the inclusion of children's rights not only in migratory or education policy, but also in all development issues and intervention strategies that affect children. There has been endorsement of discrimination of children through a culture of silencing or oversight, othering them or treating them as passive victims of development. It is important to hear from children themselves commenting on their lives, basic needs and services to inform solutions to their plights. There is also absence of children's rights-based research and policy making as well as their voices in development, education and migration despite their rights to stay with parents (article 9), best interest of the child (article 3), right to decision-making on matters affecting them (3, 12, 13 and 14), right to protection (articles 19, 32, 33, 34 and 36), as well as the right to education (articles 28 and 29) among other rights. At the moment, there are no policies on how children are left, it is assumed that parents migrate for greener pastures even where pastures are not there at all. Some parents abandon old families and begin a new life in the destination country. Children continue to be voiceless, unrecognized, discriminated and marginalised and stood for and consents made on their behalf.

Migrants' children run on parallel tracks with other children with the assumption that the playing field is level for all yet they play the same game with others on a different level field. The fields, home, school and community environments need to be reconceptualised. This calls for governments, international organisations and civil societies to advocate for children's rights, monitor and gather accurate data on the well-being of children in migrating communities. A national tracking system is needed to determine impacts of labour migration on schooling and child well-being despite their parents' migratory status. Profiles need to be developed to establish retention, performance and drop outs, movements of drop outs for wage labour or marriage. Collected data provides donors and policy makers with clear empirical and grounded framing of what children suffer from to inform solutions to get development interventions like advocacy right. Advocacy promotes awareness to reduce risks of discrimination and social exclusion of unrecognised category of left-behind migrants' children. To a greater extent, this is lacking not only in Bulilima and in Mangwe but also, in Zimbabwe in general.

There should be institutional policies and legislation relating to foster care, care centres, low-cost boarding and renting children for schooling. Policies should ensure that children have access to benefits created by migration at the same time provided with protection from the negative consequences of migration. There should be constructive policies on leaving children behind, sending of remittances, returning of parents and maintaining contact with families left behind to ease strict migration policies especially on parents and women. Right policy choices are necessary to maximise the benefits and mitigate the risks of migration. Governments limit movement or leave migrating parents with no options for visits or care for children yet governments know of economic push factors. Governments need to provide regular channels to develop work programmes and permit that allow regular return of parents.

There should also be practical-oriented intervention programmes aimed at maximising benefits and minimising the challenges of migration by parents leaving children behind. This can be done by investing on education and health for children at risks as those left behind. This can be done at national or community level. Education facilities and infrastructure can be expanded and health schemes enforced on those leaving children behind. Fees and examination subsidies, scholarships, grants and bursaries as well as free education for the under-privileged suggested by the constitution are very necessary. Low-cost intervention measures will also go a long way in alleviating the plight of migrants' children. These include health awareness campaigns and early intervention in life cycles of teenagers especially girls to learn about sexuality and the dangers of teenage pregnancies. In addition, teacher mentorship for remedial support, community-based training programmes and parent partnership in school management can be of help. Incentives for students' retention, returning to school and parent involvement can be introduced to motivate learners, teachers, parents and caregivers.

The book has social and implications in development. There should also be deconstruction of teachers, caregivers and the community's attitudes towards migrants and intolerance to other groups of people and a reconstruction of positive attitudes. Governments, NGOs and all development agencies need to deconstruct their beliefs of migrants' children as rich and well provided for. The beliefs and attitudes perpetuate not only discrimination and exclusion but also marginalisation and vulnerability. Teachers can be trained to understand emotional stress and challenges of migrants' children so that they also sensitise migrants' children on risks and challenges to promote resilience. Children can then develop survival skills, educational and social adjustments. Schools can deliver programmes on care through teachers as part of the protective mechanism for children. Caregivers also need to be trained to provide psychological support to create an educational supportive environment and to establish continuum of care when parents migrate. Teachers and caregivers can then direct also outreach of social welfare agencies to remote and highly disadvantaged areas though working in partnership with them.

Politically, the learning of physical and economic environments reflect marginalisation and lack of recognition in terms of development. About 40 years after independence, the two districts have six formal mission boarding schools and inadequate rural day schools with an average distance of more than 16–20 km in between. This reflects unequal development of regions and unequal provision of social services. The schools do not have health centres or educational supportive centres like libraries, museums, computer centres, reflecting exclusion from heritage learning. However, this is reflective of most remote rural areas of Zimbabwe. The book suggests decentralisation of social services and learning support systems by the government, NGOs and other stakeholders. Special attention needs to be paid to different provinces focusing on specific challenges of the provinces or district in relation to drop out rates, health and nutrition, maternal education, consumption poverty index and economic performance of the province. Matabeleland South province, Bulilima, and Mangwe districts' problems should be approached from the context of their geo-physical, economic, educational and political marginality.

The book also implies a turn from integrationist western intervention strategies to transformatory ones. A shift from universalistic to particularistic strategies that take into consideration the local contexts and diversity of learning environments. Equity policies in education should not universalise learners but be related to a particular group of a particular context. Western intervention equity policies are tools of marginalisation. Although poor counties cannot totally discard them, there should be caution not to rely on them ignoring local contexts.

The book has implications to migration and education policy options of partnerships for a holistic approach to address challenges and root causes. These are government, NGOs, civil society and community partnerships so that these work together in creating an environment conducive for the education of boys and girls. There are also lessons to be learnt from lack of school (teacher) and family (parents/caregiver) partnerships.

The policy strategies from the national level right to individual migrant parent, learner, teacher and caregiver should be unified and co-ordinated as suggested by the model given in this book. Fragmented strategies by individuals or institutions are not the best option.

The book also has implications for further research to influence development policy in migration and education. At present, there is scarcity of reliable data on the magnitude of the impact of migration of parents on education of their children. The number of children affected by migration is not known. Their experiences and voices remain unheard. The number of parentless due to migration is not known. Further studies are needed in this area. Performance of migrants' children compared to that of children of non-migrants' parents and the psychological effects of migration on children is neglected. There is need to reveal the magnitude of the problems affecting migrants' children by particularising them as separate category.

Conclusion

Migration is a global issue of the twenty-first century attracting unprecedented interest among the public, scholars, international organisations and policy makers. A number of studies and policies on migrants target benefits of migration through remittances. The book shifts from remittances to non-material consequences impacting on children and communities. It also shifts from general challenges to specific educational challenges and those that relate to the well-being of left-behind children. There is glaring gap in knowledge of educational and experiences of children left behind in Zimbabwe. There has been a missing link between drastic changes brought by migration in families and the damaging impact of it on the education of children left behind. If migrants' children become migrants themselves, they are likely to do as parents, repeating life cycles of children without parental care, parents who have no parenting experience giving rise to a generation of remote parenting and parentless children due to migration. With economic meltdown, unstable political situation and globalisation, migration flows will continue to increase at alarming rates in the coming decades. Abramovich et al. (2011, p. 4) predicted that if the international migration increases at the same pace, by 2050, the stock of international migrants worldwide could be as high as 405 million people. Consequently, the number of children left behind worldwide and Zimbabwe will be incredibly high. The negative impacts of migration of parents leaving children behind will soar and vulnerability of children will multiply. Issues of foster care, care centres, violation of rights, schooling challenges will even be more problematic to migration and education policy makers. Currently, the socio-economic cost of migration is not known, further studies are needed to supplement this study by quantifying the educational, economic, and psycho-social challenges to feed into policy planning. There is also need for urgent action in terms of policy framework to move away from reactionary strategies to proactive ones in dealing with migration and education. Policies should also shift from being integrationist and western to transformatory ones taking diverse contexts and capacities of the vulnerable into consideration as suggested by post-colonial theoretical framework. The people of Bulilma and Mangwe and migrants' children though vulnerable have potential for resilience through funds of knowledge acquired as they try to adapt to their geo-physical, economic, educational, cultural and political marginality.

References

Abramovich V, Cernades PC, Morclachetti A (2011) The rights of children, Youth and Women in the context of migration: Conceptual basis and principles for effective policies with human rights and gender based approach. UNICEF, New York. https://www.unicef.org/media/61041/file

Inoue et al (2015) Out-of-School Youths in Sub-Saharan Africa: A policy Perspective. WB: Washington

Josefsson J (2014) Children as the moral subjects in Ethics of migration. Proceedings from the 49th societies Ethical Annual Conference 23-26 Aug 2012. Journal Revista Ecumenical Sibiu, Vol iv: 3, 2012 pp 148–157

UNICEF (2007) Children and Migration. UNICEF. Unicef-children-and-migration-en-pdf

Index

W. Matsa, *Marginality, Migration and Education*,
https://doi.org/10.1007/978-3-030-60873-6

The manufacturer's authorised representative in the EU is Springer Nature Customer Service Centre GmbH, Europaplatz 3, 69115 Heidelberg, Germany. If you have any concerns regarding our products, please contact ProductSafety@springernature.com

Printed and bound by CPI Group (UK) Ltd, Croydon, CR0 4YY
30/09/2026
02225473-0001